Backhoe Loader Handbook
Advanced Techniques
for Operators

Backhoe Loader Handbook
Advanced Techniques for Operators

Reinar Christian

Hanley-Wood, LLC.

Copyright © 1996 Hanley-Wood, LLC.
Printed in the United States of America

Reproduction of any part of this work beyond that by the 1976 United States Copyright Act without the permission of the copyright owner is unlawful. Requests for permission or further information should be addressed to Hanley-Wood, LLC.

ISBN 0-924659-72-6

Item No. 1260

Library of Congress Cataloging-in-Publication Data

Christian, Reinar, 1962-
    Backhoe loader handbook, advanced techniques for operators / Reinar Christian.
       p.   cm.
    Includes index.
    ISBN 0-924659-72-6 (alk. paper)
     1. Backhoes. 2. Loaders (Machines) I. Title.
TA735.C53   1996
624–dc20   95-43905
                                                                         CIP

Editor: Desiree J. Hanford
Art Director: Joan E. Moran
Technical Editor: Greg Sitek
Cover Design: Steve Springer
Publisher: Mark DiCicco

Hanley-Wood, LLC. and its employees and agents are not engaged in the business of providing architectural or construction services, nor are they licensed to do so. The information in this book is intended for the use of equipment operators or those competent to evaluate its applicability to their situation, and who will accept the responsibility for the application of the information. Hanley-Wood, LLC. and the authors disclaim any and all responsibility for the application of the information.

Contents

Chapter One: Purchasing and Maintaining a Backhoe Loader

1.1	**Selecting a Backhoe Loader**	1
1.2	**Type and Size**	2
	1.2.1 Center-mount Model	2
	1.2.2 Offset Model	2
	1.2.3 Size	3
1.3	**Accessories**	4
	1.3.1 Auger	4
	1.3.2 Breaker	4
	1.3.3 Vibratory Plate	5
	1.3.4 Impact Tamp	5
	1.3.5 Sheepsfoot	5
	1.3.6 Quick-change Bucket Assembly	6
	1.3.7 4-in-1/Clamshell Loader	6
	1.3.8 Removable Loader Teeth	7
	1.3.9 Asphalt Cutter	7
	1.3.10 Lifting Forks	7
1.4	**Maintenance**	7
	1.4.1 Getting Started	7
	1.4.2 Oil	8
	1.4.3 Grease	8
1.5	**Buckets**	9
	1.5.1 Loader Buckets	9
	1.5.2 Backhoe Buckets	12
1.6	**Power Positions**	13
1.7	**Proceeding When Something is Broken**	14

Chapter Two: Safety

- 2.1 **Introduction** 17
- 2.2 **Preventable and Unpreventable Accidents** 17
  - 2.2.1 Swing Radius 18
  - 2.2.2 Operator-Laborer Relationship 18
  - 2.2.3 Cave-ins 19
    - 2.2.3.1 Contributing Factors 19
    - 2.2.3.2 Sloping vs. Stepping 20
      - 2.2.3.2.1 Sloping 20
      - 2.2.3.2.2 Benching or Stepping 21
  - 2.2.4 Taking Precautions 22
- 2.3 **Traveling Public Roads** 23
- 2.4 **Changing Buckets** 23
- 2.5 **Safety Belt** 24
- 2.6 **Eye Protection** 24

Chapter Three: Grade

- 3.1 **Communication in the Field** 25
  - 3.1.1 Hand Signals 25
- 3.2 **Reading Grade Stakes** 29
  - 3.2.1 Curb-and-gutter Grade 31
  - 3.2.2 Sidewalk Grade 32
- 3.3 **Sidewalk Grade** 33
  - 3.3.1 Hike-up 34
  - 3.3.2 Excess Material 34
  - 3.3.3 Setup 34
  - 3.3.4 Outrigger Placement 35
  - 3.3.5 Grading 36
  - 3.3.6 Obstacles 37
  - 3.3.7 Using the Loader 37
- 3.4 **Curb-and-gutter Grade** 38
  - 3.4.1 Setup 39
  - 3.4.2 Hike-down 39
  - 3.4.3 Grading 40
- 3.5 **Grade Stake Abbreviations** 42

Chapter Four: Dirt Work

- **4.1 Trenching** 45
  - 4.1.1 Cycling 45
  - 4.1.2 Digging Sequence 46
  - 4.1.3 Confinement 47
- **4.2 Straight-line Trenching** 47
- **4.3 Pushing Back** 48
- **4.4 Digging Under Utilities** 49
  - 4.4.1 Starting the Dig 49
- **4.5 Digging Under Curb-and-gutter** 52
  - 4.5.1 The Relief Hole 52
- **4.6 Footings** 53
  - 4.6.1 House Footings 53
  - 4.6.2 Line Location 54
  - 4.6.3 Adverse Material 54
  - 4.6.4 Formulating a Plan 54
  - 4.6.5 Executing the Plan 55
  - 4.6.6 Push-back/Pullout 57
- **4.7 Piers** 58
  - 4.7.1 Remove Center Material First 59
  - 4.7.2 Checking the Work 59
  - 4.7.3 Retrieving Loose Material 60

Chapter Five: Loading

- **5.1 Loading from a Pile** 61
  - 5.1.1 Spotting the Truck 62
  - 5.1.2 Loading the Truck 63
  - 5.1.3 Leveling the Load 65
  - 5.1.4 Tips 65
- **5.2 Loader Grading** 66
  - 5.2.1 Loader Bucket Edges 66
  - 5.2.2 Operator Methods 67
  - 5.2.3 Grading Guidelines 67
  - 5.2.4 Grading with a Raised Perimeter 68
  - 5.2.5 Six o'Clock 68
  - 5.2.6 Loader Floating 70

	5.2.7	Hopping	71
	5.2.8	Five o'Clock	71
	5.2.9	Three o'Clock	72
	5.2.10	Float Position	73
	5.2.11	Grading Import Material	74
5.3	**Backfilling**	75
	5.3.1	Introduction	75
	5.3.2	Compaction	75
	5.3.3	Adding Water	76
	5.3.4	Using Tire Imprints to Gauge Moisture	76
	5.3.5	Lifting	77
	5.3.6	Repetition	78
	5.3.7	Wheelrolling	78
	5.3.8	Gauging Material Flow	79
	5.3.9	Side-cutting	80
	5.3.10	Backfilling at 90 Degrees	82
	5.3.11	Backfilling at 45 Degrees	84
	5.3.12	Finishing	86

Chapter Six: Utilities

6.1	**Locating Utilities**	87
	6.1.1	Feel	87
	6.1.2	Preventing Accidents	88
	6.1.3	Utility Location	89
	6.1.4	Recognizing Trench Signs	90
	6.1.5	Watching the Spoils	90
	6.1.6	Location Technique	91
	6.1.7	X-pattern Digging	92
	6.1.8	X-pattern Variation	93
	6.1.9	Parallel Setup	94
	6.1.10	Perpendicular Setup	95
	6.1.11	Sensitivity Zone	96
	6.1.12	Looking for Existing Conduit	97
6.2	**Locating Electrical Conduit**	98
	6.2.1	Types of Conduit	98
	6.2.2	Examining Pull Boxes and Vaults	98
	6.2.3	Utility Poles	99
	6.2.4	Setup	101

6.3	Locating Water Lines............................	101
	6.3.1 Transite Concrete	102
	6.3.2 Plastic Conduit	103
	6.3.3 Steel Conduit	104
6.4	Locating Gas Lines................................	105
	6.4.1 Sand as an Indicator	105
	6.4.2 Boring	106
	6.4.3 Setup	106
6.5	Locating Telephone Lines	107
	6.5.1 Fiber Optic	107
6.6	Making Contact with Utilities	108
	6.6.1 Making Contact with Electrical Conduit	108
	6.6.1.1 Electricity Transmission.............	108
	6.6.1.2 Breaking a Line.....................	109
	6.6.2 Making Contact with Water	110
	6.6.2.1 Crimping Copper	111
	6.6.3 Making Contact with Gas Lines	112
	6.6.3.1 Making Contact in a Residential Area	113
	6.6.4 Making Contact with Telephone Lines	114

Chapter Seven: Removals and Digouts

7.1	Introduction	117
	7.1.1 Technique	117
7.2	Saving Edges......................................	118
	7.2.1 Controlled Relief...........................	118
	7.2.1.1 Example #1.........................	119
	7.2.1.2 Example #2.........................	121
7.3	Picking Up from a Flat Surface	122
7.4	Picking Up a Piece from the Grade	124
7.5	Differences in Flat Surfaces vs. the Grade	124
7.6	Picking Up Uneven Pieces.......................	126
7.7	Digouts..	127
	7.7.1 Finishing the Digout	128
7.8	Removals	129
7.9	Picking Up Individual Pieces with the Loader Bucket	130
7.10	Picking Up Individual Pieces Out of the Grade	133

Chapter Eight: Jumping the Trench

- 8.1 Front-first Jumping on the Go 135
- 8.2 Jumping a Trench at 45 Degrees 136
- 8.3 Jumping a Trench Straight On 137
- 8.4 Jumping Back First 140
- 8.5 Jumping Over an Obstacle 142
- 8.6 Jumping Obstacles 143
  - 8.6.1 High Front Tires 143
  - 8.6.2 Low Front Tires 144
  - 8.6.3 Moving the Front Tires 144

Chapter Nine: Cleaning Up

- 9.1 Introduction 145
- 9.2 Lining Up Material with a Backstop 145
- 9.3 Picking Up Material 146
- 9.4 The Pinch 146
- 9.5 Lining Up Material without a Backstop 147
- 9.6 Curb Shoe 149
- 9.7 Cleaning Curb-and-gutter 150
- 9.8 Cleaning Sidewalk 151
  - 9.8.1 Picking Up Material 152

Chapter Ten: Moving a Stuck Backhoe Loader

- 10.1 Introduction 153
- 10.2 Loader Walk 153
- 10.3 Drive-out 154
- 10.4 Pushout Method 154
- 10.5 Pivoting 155
- 10.6 Pullout Method 156
- 10.7 Walkout Method 156

Foreword

The dream machine is the backhoe loader. It has made more dreams come true than any other single piece of construction equipment. In the hands of a competent operator, it stops being a machine. It becomes the sculptor that creates the hole for a swimming pool, the foundation of a small factory, the basement of someone's home, the sidewalk through a park, a road to a secluded cabin, the landing at a lake, the trench for utility lines—each fulfilling someone's dream.

In addition to being the Leonardo Da Vinci of these activities, backhoe loaders have been the artist's knife in creating skylines, cityscapes, and pastoral scenes across the country for many years. Chances are the backhoe loader has fulfilled more individual dreams of being one's own boss than anything other than an 18-wheeler. With a backhoe loader, a person can literally get into business for himself.

Uses for the dream machine don't stop there. On one jobsite a contractor may need a backhoe, then a loader, then a fork, or an auger or maybe a broom. On the next job, the contractor may need a hammer or a small crane. Then maybe the jobs call for fine grading and truck loading, or maybe digging a trench or clearing the site. With its virtually unlimited versatility, the backhoe loader can be all these machines and more. It's the dream machine; it can do whatever the operator tells it to do.

Not too long ago, the backhoe loader, then called the tractor loader backhoe or TLB, was primarily a farm or utility tool. Most self-respecting contractors wouldn't have one on the job. It didn't take long for attitudes to change when the economy forced everyone to get more from their equipment. It wasn't practical to keep machines on the job that weren't working. The contractors that did often learned the hard way that the only good machine is a working machine.

There are only so many applications for a dozer, grader, or track loader. When the job calls for a series of machines rather than a single piece of equipment, contractors found the Swiss Army Knife of construction equipment—the backhoe loader—provided them with the answer, the right answer.

Early versions of the backhoe loader were born with manure in their tire treads, created by adding a loader attachment to the front end of an agricultural tractor and a backhoe attachment to the rear end. As the machine demand changed from the farm to the construction site, so did the design and engineering. The small utility backhoe loader became a heavy-duty workhorse capable of production trenching, high-volume backfilling, and large-capacity truck loading. With added hydraulic capacity and increased physical size, this new breed of machine easily adapted to the wide array of attachments originally designed for hydraulic excavators.

Dreams don't become reality when the backhoe loader is off-loaded at the jobsite. In the hands of an inexperienced operator, the dream potential can easily turn into a nightmare. Making dreams come true with a backhoe loader takes experience and practice.

The best way to learn all the secrets is by watching an experienced operator put the dream machine through its paces. It's a sight to behold, almost as beautiful as a choreographed ballet, seeing the experienced operator walk a machine out of a tight spot, jump a trench, fill a dump truck, fine-grade a sidewalk, and lift slabs of broken concrete or asphalt.

Everyone watching always asks, "How can they do that? How does an operator take a hulking, ugly piece of iron and turn it into a beautiful, gracefully moving dancer?"

Experience and practice are the answers. But there's more to it than that. To become the best requires knowing the machine, its capabilities and limitations, and knowing the secrets acquired over a lifetime of experiences.

In the pages of this book, the author shares these secrets with you. It took years of practice and experience to develop the skills he willingly shares with you. As you read, you'll learn the correct method of doing dozens of jobs. More importantly, you'll learn how to develop a better understanding of the backhoe loader and its almost unbelievable jobsite versatility.

The carefully selected and crafted combination of words, pictures, and illustrations will help you become a better and safer operator. You'll learn through the experiences of another, and develop a new level of confidence in your own abilities as you gain experience through the practice of what you read and learn.

This is one backhoe loader book you won't want to leave in the office. You'll want it on the machine with you, all the time. Don't settle for being just a good operator, be the best you can.

Greg Sitek
Vice President/Editorial/Publishing Services
Randall Publishing Co. Inc.

Acknowledgments

Thank you to some special people who have in one way or another helped me in the production of this book:

Kyle and Trenton: I thank you both for being my best friends, my favorite playmates, my sons.

Sandi: Thank you for picking up the slack, for being my support, and for being the one I look forward to growing old with.

Ray and Shirley Rucker: If it wasn't for Rucker Backhoe, your patience, and your ability to teach, none of this would have ever transpired. Thank you both.

Dr. James Christian: For allowing me to steal you away from your writing to help me with mine, time after time, thanks dad.

Sheldon: Thank you for the type of help only a true friend can give.

David Sheridan: Thank you for your photographic ability, equipment, and for your friendship.

Johnny Hamilton: Thank you for opening doors that would have otherwise remained closed.

Chapter One: Purchasing and Maintaining a Backhoe Loader

1.1 Selecting a Backhoe Loader

Maintaining a backhoe loader doesn't start with, and is not limited to, an oil change and a couple of tubes of grease. It starts with the selection of a piece of equipment that, considering size and work requirements, is best suited to someone's needs. Once a backhoe loader has been purchased and attachments added, a maintenance program should begin.

Which is the best backhoe loader on the market, and what is the best size? All backhoe loader manufacturers produce a good product; if they didn't, they wouldn't be in business. Currently there are several major backhoe loader manufacturers, each of which offers a full line of machines in various sizes. The brands differ in significant ways because manufacturers try to appeal to different aspects of the construction industry. What separates one backhoe loader from another is specialization and application. For example, on one backhoe loader, the backhoe has a heavier, more curvaceous boom designed to achieve greater digging or breakout force. In achieving this, the overall machine suffers from a slightly higher center of gravity. Another backhoe loader has a low-flow, high-pressure main hydraulic pump to achieve smoother valve response, but it sacrifices cycle speed. Another backhoe loader has a high-volume, high-pressure main pump that is capable of a rapid cycle speed, but this makes smooth control a little more difficult on the operator.

A successful financial decision starts with the purchase of the right piece of equipment for the right job.

This leaves the purchaser with the question of application. What exactly will be done with the backhoe loader? What jobs will it perform? Will it be a rental machine that will be doing a little of everything, or will it be a backhoe loader that digs water and sewer laterals all day on

Chapter 1: Purchasing and Maintaining a Backhoe Loader

a large housing project? Will the backhoe loader belong to a paving contractor who will use it only to remove asphalt and concrete?

Ideally, money should not be a factor when purchasing a backhoe loader. They are expensive, but what the purchaser does not pay for now, he or she will pay for later. A buyer knows exactly why the backhoe loader is being purchased; sticking with that reason is the best idea. A contractor's primary concern is to buy the best machine for the job, because then the backhoe loader will bring an excellent return on the investment, provided it can fulfill two requirements: The operator should be able to use it to its greatest potential, and the purchaser should have a mechanic who can properly maintain it.

1.2 Type and Size

All backhoe loaders are capable of performing the same operations. (The only exceptions are depth reach and breakout force.) But some backhoe loaders may perform tasks a little easier or a little quicker than others. As an example, two major types of backhoe loaders are illustrated in this chapter to show the differences as they pertain to the proper type of backhoe loader and the correct size of a backhoe loader.

1.2.1 Center-mount Model

With a center-mount model, the backhoe section of the machine is in a permanent, stationary position at the rear center. The most widely used backhoe, a center-mount is an all-purpose design, performing excellently in most applications.

The center-mount is less expensive, provides better balance, has fewer maintenance problems, and is easier to set up on most jobs. In short, it offers the best overall value for most applications.

There are a few disadvantages to center-mounts. Because of the outrigger configuration, it is not as stable as an offset model is in critical footing situations, and digging parallel to existing structures requires more time and operator skill than doing so with an offset model. For example, when digging next to an existing wall, the backhoe loader must back up to the wall (with the boom swung over to avoid contact) at an 80- or 90-degree angle, make a pass, then drive forward and over in order to back up 10 or 15 feet for the following pass.

1.2.2 Offset Model

Whether by hydraulic control or by loosening holding bolts, an offset backhoe has the capability of moving along a slide rail to either extreme side at the rear section of the machine. Being able to shift the backhoe section from one side to the other makes this type of backhoe loader a favorite among block masons because of their frequent need to dig close to existing structures.

Because the outriggers are located on either end of the slide rail and therefore provide an extremely stable vertical environment, offset backhoe loaders are known for their excellence in footing and dirt work.

Drawbacks include the extra weight needed to provide the tractor with the backhoe slide assembly. Its heavy weight produces a very light front end when traveling on any sort of incline. The offset model also requires quite a bit more skill from the operator in fine grading situations with the loader.

1.2.3 Size

Because of a backhoe loader's versatility, one size can perform a wide spectrum of jobs; it is not necessary to have a different size machine for each individual job. While a purchaser does not want to damage a smaller backhoe loader by pushing it just to maintain what a larger backhoe loader could do at normal operating speed, a purchaser also does not want a larger machine doing a job that a smaller one is better suited for. Purchasing the biggest backhoe loader in the series to try to accomplish all size jobs more easily while—still making the machine last longer—is not possible. There are other variables that must be considered.

First, a small backhoe loader doing the job that a large backhoe loader should be doing is working much harder than it should. The engine is turning higher r.p.m. so that the main pump can supply the system with a constant hydraulic flow. This, in turn, creates higher engine and hydraulic oil temperatures, leading to more frequent oil and filter changes.

Second, a backhoe loader constantly pushing its maximum weight-lifting limit will have a much faster cylinder packing fatigue or failure. A larger machine being operated under normal conditions will be able to follow the manufacturer's recommended periodic servicing tables. Being able to follow these recommended practices will lead to less frequent and less expensive repairs.

A third factor is the sheer size of a large backhoe loader relative to the job it is performing. For example, there are a couple of different, larger machines whose backhoe booms are wider than the width of a 12-inch bucket. Long, custom-built, 12-inch buckets are available for these machines, but these still will not allow the backhoe to dig deeper than the bucket pins, thus eliminating them from certain jobs.

Fourth, a large backhoe loader weighs more than its smaller and lighter counterpart. When a backhoe loader is in the digging position, the outriggers are supporting most of the machine weight. The operator can place the outriggers of a smaller backhoe loader on the sidewalk, without the danger of cracking it. The larger backhoe loader may have a problem.

Chapter 1: Purchasing and Maintaining a Backhoe Loader

Finally, maneuverability in tight quarters is compounded exponentially when using a large machine.

1.3 Accessories

There are as many accessories for backhoe loaders as there are machines. Attachments are available to do everything from mowing the yard to sweeping the street. The focus here is on the most widely used accessories.

1.3.1 Auger

Like many attachments, an auger is a hydraulically operated device parasitic to the machine's hydraulic system. It generally uses the bucket or extendable dipperstick hydraulic circuit to power a gear driven motor which, in turn, is used to drill holes in the ground.

Auger bits range in size from a diameter of 6 to 48 inches and, with the addition of extension hoses and extension bits, depths of up to 20 feet can be attained.

1.3.2 Breaker

Using the boom for down pressure, a breaker feeds off either an aux-

Augering a streetlight foundation.

iliary hydraulic system or the extendable dipperstick system to drive an internal concentric weight that is impacted against an external striker plate. The result is enough impact force to shatter concrete by means of a removable bit or point at the end of the breaker.

Impact breakers are available in different sizes to accommodate all sizes of backhoe loaders. The smaller breakers, used for residential-sized machines, are ideal for breaking concrete slabs and for preparation of sidewalk and street removal.

1.3.3 Vibratory Plate

A bucket cylinder is needed for control on a vibratory plate. Either an auxiliary circuit needs to be installed on the backhoe loader or an extendable dipperstick circuit can be used to run the device. When the latter of the two is chosen, a great deal of reach is lost, compounding the time needed to complete larger projects.

A vibratory plate works best with the compaction of lighter materials. The backhoe loader hydraulic system feeds a concentric pump that vibrates a large shoe on the bottom of the plate. Compaction occurs by placing the shoe in the desired area, booming down to apply pressure, and opening the hydraulic circuit, thereby creating a high-speed vibration against the ground.

The incorporation of the bucket cylinder changes the angle of the shoe relative to the ground, making the device extremely adept for flat or slope work.

1.3.4 Impact Tamp

An impact tamp is a breaker motor with a shoe attachment that enables the tamp to serve a dual purpose. On harder materials such as clays or adobes, where a vibratory plate and its vibrating shoe would be useless, an impact tamper can be used to break up the material by hitting or impacting it.

Like a vibratory plate, the impact tamper is run off either an auxiliary or extendable dipperstick circuit, keeping the bucket cylinder free to maneuver.

1.3.5 Sheepsfoot

Depending upon the size, a sheepsfoot roller can have as many as eight rows of spiked steel wheels connected at the centers, with a gap between each wheel. A bearing assembly connects the compaction roller to a frame, to the backhoe, and to points at the boom end and bucket cylinder. The name *sheepsfoot* comes from the pattern left by the spiked wheels that are fitted with removable shoes.

This type of compactor is ideal for longer trenches where speed and maneuverability are musts. As the compactor is wheeled forward and backward on the material, the backhoe's dead weight is used for the

Chapter 1: Purchasing and Maintaining a Backhoe Loader

Sheepsfoot compactor wheel.

down force. The framework necessary to support the weight of the backhoe allows the operator to use the sheepsfoot as the bucket would be used, pushing and pulling the backhoe to advance the action. Change over speed is as quick as changing a bucket, and maintenance is a matter of making sure the sheepsfoot is greased every few hours of operation.

1.3.6 Quick-change Bucket Assembly

During the excavation of a house foundation, an operator will probably change the bucket as many as 10 times to accommodate the different sizes and types of footings. This is where a quick-change assembly is helpful. Although it does not take long to change a bucket with a normal pin setup, the quick-change system saves the operator from having to move back and forth between the bucket on the ground and the backhoe levers in an effort to align the bucket pins. One bucket pin is pulled and the special yoke is allowed to unhook from the bucket. The operator then needs to swing over, hook the new bucket, and reinsert the bucket pin. This system eliminates the trial and error of pin alignment, thereby increasing bucket changing speed.

The disadvantage is that a full set of new buckets must be purchased in order to interchange with the new system.

1.3.7 4-in-1/Clamshell Loader

A clamshell loader bucket was designed to help an operator in the following instances. A trash pile is loaded into a truck, and because the material is light, everything flops out of the loader bucket back onto the ground. Or a pile of leftover material in the street is picked up, but some of it has to be shoveled because there is no backstop to use.

A clamshell loader bucket is split lengthwise and hinged on top. A couple of cutting edges are added to the bottom, a couple of small hydraulic rams are attached to the sides, and the unit is patched into the hydraulic system.

The advantage to this loader bucket is its ability to grasp an object. Through the activation of two small rams placed on either side of the loader bucket, the front half of the loader is able to be raised, allowing a pinch point in which to pick up an object or extra spoils.

Grading is made a little easier because the operator has three cutting edges to choose from instead of just one.

The operator must use a great deal of care when lifting or manipulating heavy objects because although there is the benefit of splitting the loader in half to gain its grasp, there is also a weakening of the front loader half, particularly the lower front cutting edge.

1.3.8 Removable Loader Teeth

Although it takes a few minutes to change them, removable loader teeth are well worth the investment. They are used to remove large sections of thin, uncut asphalt. Bucket teeth make penetrating easier and faster, and they are an asset when removing asphalt or thin layers of concrete. Removable teeth give the operator greater versatility on the job.

1.3.9 Asphalt Cutter

An asphalt cutter is a sturdy oversized pizza cutter that bolts to the front underside of a loader bucket and uses the front-end weight of the backhoe loader as its down pressure. It is not the best tool available for cutting asphalt, but for small jobs in situations where nothing else is available, it can be a plus.

1.3.10 Lifting Forks

Lifting forks are ideal when an operator is trying to balance a 20-foot length of 12-inch plastic conduit on the loader. A series of hooks or holes are welded or drilled at the top of the loader and a steel rod slides through them. Forks hang from the rod; changeover time and maintenance are minimal. Lifting forks are an absolute must for pipeline use.

1.4 Maintenance

1.4.1 Getting Started

A backhoe loader is not a person, but it goes through many of the same early-morning traumas that a person does and for the same rea-

Chapter 1: Purchasing and Maintaining a Backhoe Loader

sons: A backhoe loader, like a human, has a heart, limbs, and a circulatory system.

The backhoe loader's engine oil should be checked before it is started every day. A radiator fluid check a couple of times a week is sufficient.

After the engine is warmed up, the hydraulic oils checked, and the tractor greased, the operator should raise the loader just above the ground. The boom should also be raised if it was left lowered to the ground overnight. With the weight off the ground, the operator should check the tires and finish greasing the points where movement of the tractor was necessary.

1.4.2 Oil

A backhoe loader engine's heart is its oil pump, and its blood is its oil. In the morning, before the engine is started, the oil is in the bottom of the pan, cool and thick from sitting all night.

The engine's oil does four things: It lubricates, cools, cleans, and acts as an internal cushion. It cannot perform any of these tasks until the oil pump is engaged, and that can't happen until the engine is started. Yet to prevent damage, the engine needs the lubrication and cushioning by the oil before the engine is started.

A two minute cool-down period is mandatory for backhoe loaders equipped with turbochargers.

How does an operator lubricate and cushion a dry engine long enough on start-up for the oil pump to pick up the oil from the bottom of the pan and circulate it through the engine? It's not possible. Every morning an operator is causing damage to the engine by starting it, and there is nothing that an operator can do about it. It is one element of maintenance that an operator must adjust to. The oil carries and discards the minute bits of metal that have been ground off the internal surfaces of the motor. As the filter becomes clogged, more and more pieces of this metal and various bits of engine debris are allowed to recirculate throughout the engine, compounding damage.

Frequent oil changes are invaluable. The more engine junk that can be prevented from circulating through the engine, the longer it will last.

The oil should be checked first thing in the morning. After starting it, the machine should be idle for at least 10 minutes. Also, when any of the hydraulic controls are activated while the tractor is running, it makes the engine work harder, so the operator should hold off pulling any levers during the 10-minute warm-up period.

1.4.3 Grease

A perfect time to grease the backhoe loader is while it is idling. Squeaks, hot spots, and gaulding are the order in which bushings and pins become damaged when the parts are not greased. Once a pin or bushing becomes gaulded, it acts like sandpaper against its counterpart. The object is to keep the pins and bushings lubricated so gaulding doesn't occur.

The tolerance between a well-fitted pin and bushing is almost immeasurable, perhaps .002 of an inch. When grease is injected into the bushing by means of the grease zerk, that thousandths of an inch space is filled with grease, creating a cushion between the pin and the bushing. The pin then floats in the grease, never touching the sides of the bushing. Squeaking occurs when the grease film becomes so thin that the pin comes in contact with the bushing wall.

> *A pin should be replaced as soon as it becomes gaulded.*

When this occurs, it is a forewarning to disaster because the resulting friction produces heat on the surface of the pin. The heat melts or washes out the remaining grease from the pin's surface, and soon the two are rubbing together metal to metal, producing hot spots.

Hot spots are minute flares that occur on the surfaces of the pin or bushing. These sand-particle-sized flares range from 600° to 1,200° F, and during the split-second they are present, they create tiny pits. The pits themselves are not the cause of the damage but instead it is the metal fragments from the pits, carried around the pin, that scar its surfaces. This is called gaulding.

Once a pin is gaulded, it will no longer hold grease because the tolerance between the pin and bushing is too large. When this happens, the pin makes noise and, as time progresses, permanent damage will occur.

1.5 Buckets

A backhoe loader maintenance program should not be limited to the machine and its operations. A maintenance program should include all the running gears, systems, buckets, and attachments.

Maintenance of these items needs to match that of the rest of the machine, with the addition of one item: prevention. When a backhoe bucket becomes twisted or bent, or a loader bucket starts to bow in the center, they are permanently damaged. Once the metal has become distorted, it can never be brought back to its original integrity. Every precaution must be taken to preserve the buckets' original forms.

1.5.1 Loader Buckets

Chances are an operator will damage the loader bucket one of two ways: The corner will be used to pry something that is too heavy, or it will be used to lift something close in, with the straps or chains improperly placed.

The following is an example of an obstruction that needs to be removed from a large street grading project, and the method an opera-

Chapter 1: Purchasing and Maintaining a Backhoe Loader

tor could use to remove it.

The first thing the operator should do is set up the backhoe and the bucket to determine the perimeter of the structure. Once the size is established, the operator would try to pry the concrete with the backhoe bucket. If that doesn't work, the operator might try to use the loader, because if the operator can move the piece, it may be possible to lift the piece.

Because the piece is square, the operator would approach it at one of its corners with the loader. It is easier to monitor results if the operator can watch the loader work the piece, so the operator would make sure the corner of the piece is off to the side of the loader that can be seen.

In about a half-open position, the operator would bury the loader bucket into the ground, catching the corner of the structure just below ground level. With a good bite, the operator would put the machine in low gear and slowly ease back on the loader control while applying forward pressure with hopes of rocking the piece. If it still won't budge, the operator would pull back hard on the loader control and work the transmission to rock the machine even harder against the piece of concrete. If the piece starts to move, the operator may notice that the harder the machine is rocked, the more traction is lost as the rear side opposite the loader starts to lift off the ground under the immense loader pressure. The solution the operator may think is to swing the boom to the light side of the tractor. The rear may still be light, but at least it's light on both sides.

> At the end of the day, the operator should lower the backhoe and loader buckets to the ground.

With the piece starting to lift, the operator can get a little more leverage by incorporating the curl of the loader. With the rear of the tractor completely off the ground and the front tires squatting almost to the rims, the operator has achieved the goal: The piece has completely broken loose, and the operator can now get a chain around it to lift it.

The operator would back off the piece and set the loader down to secure a chain. But what if, as the operator brings the loader to the level position on the ground, the operator notices that the left side of the loader hit the ground first? Fortunately for the operator, none of the tires on that side are flat.

After getting off the machine to take a closer look at the loader bucket, the operator sees that the left side of the loader bucket is bent. But because there is nothing the operator can do about it right now, the operator continues to pick up the piece by setting up to hoist the chunk of concrete from the grade.

This loader bucket is equipped with hook loops on either top edge at the corners. The laborer has set up a couple of chains in a basket under and around the piece, with a single length of chain leading

Buckets

from the middle to which the operator can hook the loader bucket.

But the length of chain isn't long enough to reach both loader loops, and the operator is wary about lifting all that weight with just one side of the loader. The operator decides to run the chain in a circle through the loader loops and tie it back upon itself, making a secure point in the middle of the loader where the operator can hook the basket chain and lift while still spreading the weight to both sides of the loader.

The operator should then ease the tractor to the edge of the piece and lower the loader. After the laborer hooks up the chain and gets away from the pulling point, the operator would slowly apply loader pressure to the chain until the slack is taken up and then push the loader cylinders to their limits. The piece starts to lift, but what if it's still too heavy to lift out of the hole? The operator would lower the loader down to the ground while still keeping the chain taut. With the loader bucket a few inches above the ground, the operator then decides to back up and drag the piece out of the hole. As the piece starts to move, the operator backs up more.

With one chain leading down the center of the loader bucket, the operator is able to pull relatively straight. As the piece nears the top of the hole, there is a little more resistance and the rear tires start to spin. The operator decides to drop the boom to get more weight to the rear. As the piece crests the edge, the rear tires start to break loose again. With the piece almost over the edge, the operator curls the loader bucket, trying to get the piece just a little farther. Finally, the piece comes out of the hole.

But while getting off the backhoe loader to unhook the chain, the operator notices that where the single chain ran down the center of the loader bucket, there is now a crease. As the operator debates how to tell supervisors about the crease, a supervisor looks at the damage and tells the operator that a heavy bulldozer could have easily removed the piece.

What could have been done to avoid the destruction of the loader bucket while still getting the job accomplished? Quite possibly nothing because the job was too big for the piece of equipment. The operator could not have known that until the project was started. But there were signs along the way that, if heeded, would have pointed toward a bad ending. Paying attention to such signs is an example of preventive maintenance.

At the very beginning, when the operator scratched the surface to find the perimeter of the obstruction, the operator should have taken it one step further. After establishing the perimeter, the next step should have been to dig along the sides down and to the bottom of the piece. The operator could have then seen that the piece was too large for the machine.

11

Chapter 1: Purchasing and Maintaining a Backhoe Loader

If the operator still thought the piece could have been moved even after seeing its enormous size, preventive measures could have been taken. First, the operator should never apply serious pressure to just one corner of the loader bucket. When the operator must push against something, it should be done with the entire loader edge making full contact with the piece. When the piece is smaller than the loader bucket, the operator should center the piece in the middle of the bucket and push. Using the bucket as a pry bar or banging it against something to break it loose is pointless.

Second, the rocking of the piece should have been done with the backhoe when the operator was already set up for digging along the sides. The operator would have tried to move it with the backhoe and could have realized that the piece was too big for the machine.

Finally, having the laborer make a basket with the chains around the piece while the operator set the chains on the loader bucket was a major mistake. The chain loops are on the ends of the loader for a reason; the manufacturers of the backhoe loader determined the stress points of the loader. When the operator set up the chain in a loop and then ran the single chain down the center of the loader bucket, the operator placed all the weight on the weakest portion of the loader: the top center. When lifting with the loader, the operator should always do so with the weight distributed equally on each end.

Of all the mistakes made in this scenario, the last one was the worst because it has the potential of being fatal. An operator should *never* use chains or straps attached to the loader to pull or drag an object toward the backhoe. If the chain or strap breaks while the operator is pulling directly toward the backhoe, the operator is in a fatal position. If the loader cannot lift the object vertically, or very close to it, the operator is using the wrong piece of equipment.

1.5.2 Backhoe Buckets

Concrete and heavy asphalt removals are probably the biggest cause of bent and twisted backhoe buckets. The weight or size of the concrete or asphalt is not what bends a bucket. What bends it is the improper positioning of the piece in the bucket as the operator applies pressure, and a piece that is too large relative to the type of force the operator is applying to the bucket.

For example, the operator is in the middle of removing a section, and trucks are lined up to the left. The section consists of 6-inch-thick concrete in a stretch 12 feet wide by 25 feet long. A stomper gave the operator a good starting point, but then the stomper's cable snapped and the stomper is down for the day. The operator knows that the section is too thick for the backhoe without proper stomping, but the trucks are already there and the work needs to get done.

The operator reaches out and wedges the teeth in the area broken

Power Positions

by the stomper and applies boom pressure, but nothing happens. To increase leverage, the operator pulls a little closer to the starting point and rolls the bucket back, with the tips of the teeth under the edge. The piece starts to crack.

The operator wedges the bucket a little farther under the piece and rolls the bucket back even harder, but nothing happens. The concrete is in too much of a bind to release, so a relief point must be found.

The operator quickly repositions the backhoe at a slight offset angle so that a tooth can get under the edge of the concrete and break a smaller piece, creating a relief point. The operator wedges the corner of the bucket under the edge and rolls the bucket back. The operator works with the bucket and the crack increases, eventually leading to a broken section lifting from the center of the slab. A relief point has been made and the removals can continue.

> *The strongest part of the backhoe bucket is in the corner where the side meets the cutting edge.*

But when the operator repositions back to the center of the section, the bucket is frowning. The operator has to spend the rest of the day dealing with the fact that one side of the bucket is hitting the ground before the other.

There is an easy solution to this scenario: A larger backhoe loader may have been able to relieve the concrete without inflicting damage. The operator had no business trying to tackle the job without the means to relieve the slab properly.

The one smart step in this situation was when the operator realized that the strongest part of the bucket was in the corner; that was the part the operator used to apply all of the force. Had the operator tried to wedge a corner of the slab in the center of the bucket, the distortion to the bucket would have been severe enough to shut the backhoe down until the operator changed buckets.

There are different types of force an operator can apply to the bucket. These forces are determined by the combination of any two or more of the following four factors: the position of the machine relative to the removal piece; the curl of the bucket; drawing in of the crowd arm; and the action of booming up or down.

1.6 Power Positions

The following analogy helps explain the positions of force. If someone holds a 1-gallon bucket filled to the brim with water straight out at arm's length, it doesn't take long before the bucket starts to lower back to the person's side. It's much easier to hold the same bucket at about half that same distance with the elbow bent.

A backhoe's power positions work the same way. The farther an operator reaches, or the higher up the operator extends, the more the

Chapter 1: Purchasing and Maintaining a Backhoe Loader

backhoe's power capability is reduced. The closer the operator keeps the boom and crowd arm to the backhoe while working, the more that power is increased.

Because of these differences in power, positioning the backhoe relative to the piece to be removed needs careful attention. There is no reason to put the extra strain on the backhoe by lifting heavy pieces that are way out there when it is just as easy to get those pieces working close in or by pulling the backhoe closer to the pieces.

Incorporate the bucket while the boom is low to the ground and the crowd arm is in tight to the backhoe and the operator will see a dramatic increase of breakout force. This is when buckets start bending.

The bucket really begins to break when the operator brings in the boom. With the tips of the teeth under a piece of concrete waiting to be broken, and the rest of the bucket lying as flat on the ground as possible, the operator should boom down until the weight of the backhoe is on the bucket, and then slowly crowd out while rolling the bucket on its back. This is the most powerful bucket position the backhoe can achieve with the bucket on the ground. If this position must be used to break out concrete or asphalt, the operator should change to a much stronger 1-foot bucket. Although 1-foot buckets are much stronger than their larger counterparts, under extreme pressure they have been known to buckle behind the front cutting edge.

These examples are not intended to show how to get more force to the ground or to show how to break out thicker pieces of concrete or asphalt. They are included in hopes of clarifying why a particular bucket bent in the past, and showing how to avoid potential problems by choosing the right size bucket for the job.

1.7 Proceeding When Something is Broken

The most significant criterion when deciding the most important parts of a backhoe depends on whether the backhoe can still do its job while the item is broken or damaged. In some instances, taking the time to fix or repair an item must take a back seat to completing the job.

For example, hydraulic lines can leak or break. If a line is broken and is dumping fluid rapidly, the job must stop and the line repaired. If a specific line is unavailable, any other spare hose that will match it can be used until the correct hose can be installed.

A backhoe loader is equipped with a hydraulic fluid reservoir. This reservoir contains gallons of extra fluid that can be used in case of a leak. Repairing a line with a slow leak can be put off by adding fluid as necessary.

Cavitation is the implosion of air bubbles in the fluid which can result in burning or eroding the metal surfaces in the main hydraulic pump. An operator must be very careful to maintain a sufficient sup-

Proceeding When Something is Broken

ply of oil in the fluid reservoir or else cavitation can occur, slowly destroying the pump. An operator should keep this in mind when adding fluid because of a leak.

In another example, a backhoe loader was parked overnight on the jobsite. The operator notices that the front tire is flat. Since a spare is not available, the operator informs the foreman that the machine is down until the tire repair people come and fix it.

Depending on what was planned for the day, the work may still be able to be done. If the operator was going to road the backhoe loader five miles before working, then the work will not be done. But if the operator was going to travel a couple hundred feet down the road to start removals or trenching, then the machine isn't down, it's just inconvenienced.

The backhoe loader has two steering systems: one via the steering wheel, the other via the steering brakes. With the loader bucket down low enough to raise the front tires off the ground and ride them on the lip of the cutting edge, the operator should engage the reverser and travel backward to the location, using the steering brakes to maneuver the tractor. Until the tire is fixed, the operator should keep the loader bucket down.

When the time comes to push back, the operator should either set the loader bucket at a level position to the ground and push or, with the tires still off the ground and the loader bucket in the same level position, lift the outriggers, turn around, and drive forward, sliding along on the lip of the loader bucket.

Chapter Two: Safety

2.1 Introduction

Compromises can usually be found on a jobsite, but compromising a worker's safety is never negotiable. Job safety is always the first priority. The job is second. A good backhoe loader operator knows what movements the machine can and cannot safely perform and is comfortable suggesting options and giving directives rather than running the risk of injury.

> **SAFETY TIP**
> Safety starts with good communication.

Every decision connected with a backhoe loader's operation is the responsibility of the operator. Completing the job properly is a serious matter and, in order to take the task seriously, the operator needs to be able to stay focused and concentrate.

This is a tall order, however, and accidents will inevitably occur. For example, operators can get distracted by rehashing an argument from last night, by pondering the next vacation, or by pushing a little too far when trying to impress a supervisor or making a deadline. These and other instances can interfere with concentration and contribute to unsafe working conditions.

2.2 Preventable and Unpreventable Accidents

There is a difference between a preventable accident and an unpreventable accident, even though the outcome is the same: What happened was not intended.

The difference has to do with cause. In an unpreventable accident, such as the backhoe being hit by a meteorite, there is nothing the operator can do to stop the occurrence. In a preventable accident, such as injury to another worker who was too close to the backhoe while it

Chapter Two: Safety

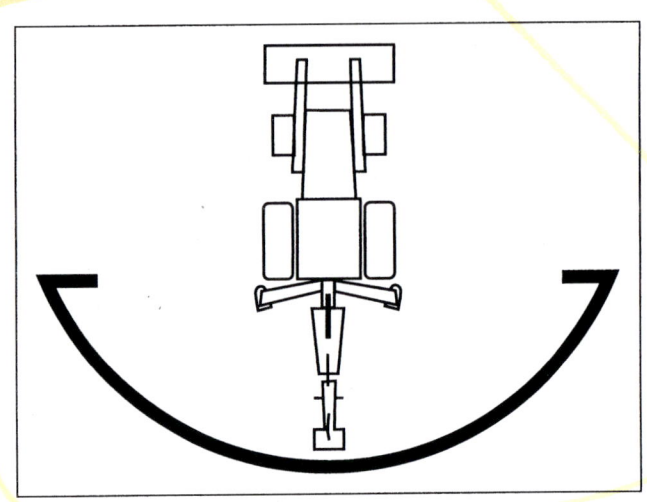

An operator must be constantly aware of everything inside the swing radius.

was in use, the operator could have taken steps that might have kept the accident from occurring.

2.2.1 Swing Radius

The swing radius of a backhoe is a prime area for accidents to occur. Ideally, the backhoe's swing radius should remain safely free of all ground traffic because this would eliminate the danger of accidents. Unfortunately, this isn't feasible in real-working situations on the jobsite. There are trenches that need to be grade-checked, small piles of rubble that must be retrieved during removals, and pipes that need to be steadied as they are laid in the trench. For these and many more reasons, laborers must often remain in the swing radius of a backhoe while it is being operated.

For example, an operator is loading trucks with broken concrete. One of the pieces falls to the ground, sending a concrete splinter into the arm of a worker. By any definition, this is an accident: The operator did not *intend* for it to happen.

But the accident was also preventable. What was the laborer doing so close to the truck? If the worker had been positioned outside the swing radius of the backhoe, or if the operator would have asked the worker to move a safe distance, the accident would have been avoided.

2.2.2 Operator-Laborer Relationship

Accidents can also happen because of assumptions. Many operators work with the same laborers every day. The laborer usually works close to the backhoe because the operator handles the backhoe well. The operator is comfortable with the laborer because the laborer is conscientious when working close to the backhoe.

However, an operator can become too comfortable with the laborer and vice versa. Many laborers become careless when working around backhoes, and they don't give a second thought when the backhoe bucket swings by, a foot away from their head. This can lead to a serious, yet preventable, accident.

2.2.3 Cave-ins

Cave-ins, although preventable, are one of the leading causes of death in excavation. When a section of dirt is removed from an otherwise solid surface, the area surrounding the void becomes unstable and unsafe, and it creates the potential for a deadly trap.

An operator's responsibility does not stop at the tips of the backhoe bucket; it encompasses the entire excavation area, including the digging area, stockpiles, travel areas, and any point where personnel can come in contact with the backhoe. Different materials hold a better vertical wall than others, but all vertical walls that may present a hazard must be sectioned off to keep personnel at a safe distance, shored to preserve vertical integrity, and the sides stepped or sloped to remove the load from the sidewalls.

In addition, counties and states have different safety codes regarding sloping, benching (or stepping), and shoring. Most safety agencies require sloping, benching, or shoring if a trench is over 4 feet deep and if personnel are required to be in the trench.

2.2.3.1 Contributing Factors

A few factors contribute to cave-ins:

1. Dig level. This refers to the sides of the trench, not the bottom grade. The operator should make sure the backhoe loader is in a level position before starting to dig. If a trench is started when the backhoe loader is in even the slightest off-level position, an overhang will occur on one side of the trench. Allowing an overhang is the same as undermining a ledge: The dirt under the top material is the support, and if the dirt is removed, the ledge will fall.

2. Placement of spoil material. When placing spoil material, two rules should be applied. First, the operator should never place the spoil material on the immediate edge of the trench. This places unnecessary weight on an already unstable surface. Second, the spoil material should be placed far enough away from the edge so that if moving or pulling the material becomes necessary, the backhoe loader's weight is away from the sides of the trench. An operator should never push back with the bucket on the top edge of the trench. If the trench's depth does not allow the operator to use the bottom to push back, the operator should elevate the outriggers and the loader bucket and slowly drive forward to the next pass.

Chapter Two: Safety

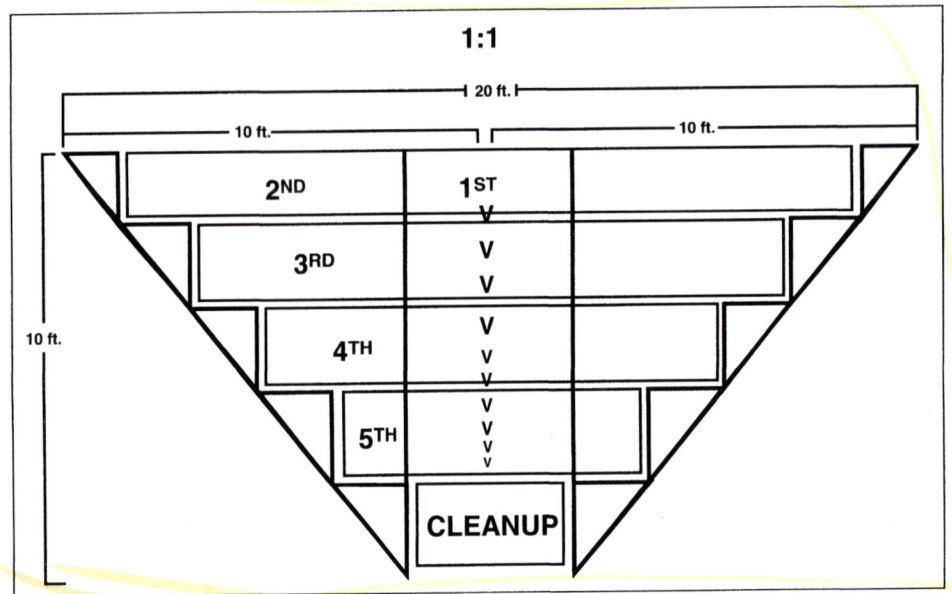

A typical step-digging sequence.

3. Other equipment working on the same project. Unless kept at a safe distance, any type of equipment may be detrimental to a trench because of size, weight, or vibration. For example, large vibratory rollers that compact the subgrade and rock-based layers in street sections can transmit enough shock waves through the ground to cave in trenches from 100 feet away. Impact tampers and breakers used for demolishing streets and compacting soil can also cause problems because they can transmit vibrations a considerable distance, depending on the soil condition.

2.2.3.2 Sloping vs. Stepping

Sloping, benching, or stepping the sides of a trench serve the same purpose, preventing the sides of the trench from caving in. Material type is probably the best guideline to use when determining whether to slope or bench. Benching entails far less moves than sloping and is therefore faster. However some material, like sand, won't hold a solid bench. Whether or not material can hold a bench can only be determined when the process begins.

2.2.3.2.1 Sloping

Sloping creates a smooth, constant slant on the side of the trench from the top edge down to the bottom corner of that same side. Sloping is mandatory when personnel must work at the bottom of a sandy trench when shoring is not an option.

The ratio 2:1 means that for every 1 foot of vertical elevation, the

Preventable and Unpreventable Accidents

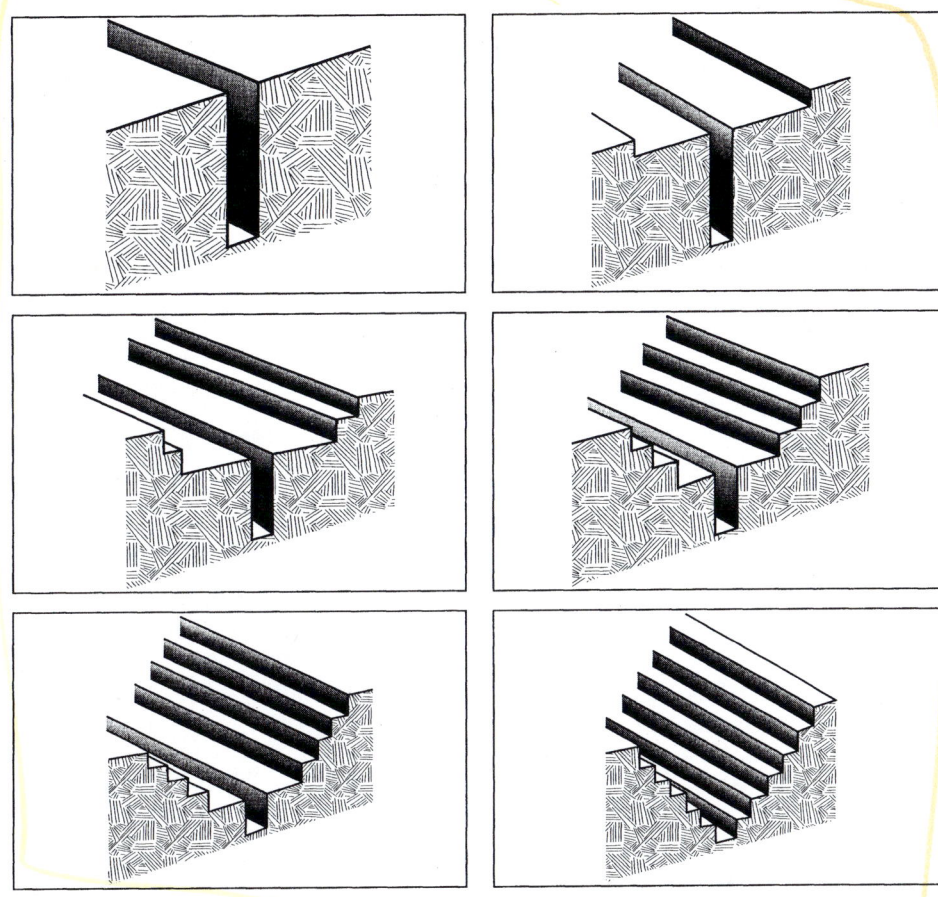

Care should be taken to make each step clean and square.

grade at that 1 foot elevation is adjusted 2 feet on a horizontal plane. On a side profile of stairs, for example, each tread, or step, has a front (vertical) and a top (horizontal). The front of the step has a rise of 1 foot and the top of the step has a depth of 2 feet. If a line was drawn from the nose of the first step, where the front of the step and the top of the step meet, to the nose on the last step, the line would be the slope.

Once the instruction for the amount of slope has been given (1:1, 2:1, 3:1, etc.), the operator should set up on to the trench as normal and proceed to dig the trench, stopping just shy of the desired bottom grade.

2.2.3.2.2 Benching or Stepping

When benching the sides of a trench, it is important for the operator to keep a visual picture of what is trying to be accomplished: the profile of a staircase.

Chapter Two: Safety

If the material is soft enough, the first bench on each side of the main trench can be dug from the original centerline position. The operator should position the bucket in the center of the first step and pull in, maintaining a parallel line to the existing trench.

The operator should dig the first bench all the way down to the top of the first step on both sides. As the passes are being cut, the loose dirt will fall in the existing trench. It is necessary to make passes on the original centerline trench and remove the spoil material. The operator should never let the original trench fill more than half way with spoil from the steps.

When the first step on each side of the main trench is complete, the operator should position the backhoe and begin digging the second step. As each of the steps are being cut, it will be necessary for the operator to clean the completed steps because of falling dirt from the preceding step.

When each side of the trench has been benched, the operator should work from the bottom (when possible) and use the swing in conjunction with the raising of the boom to knock the points off of the benches. The operator should continue to smooth out each side of the trench until the desired effect is reached. A couple of passes in the bottom of the trench can clean the loose material and make finish grade.

Operators should keep something in mind when stepping. If, for example, 4 feet is the maximum depth a trench can attain without shoring or sloping, 4 feet is not necessarily describing a specific path of a trench. Instead, that measurement is the maximum amount of vertical material allowed in a given area. Simply stated, if 4 feet is the safe limit for a trench, then the face of any step must not exceed 4 feet, and the top of any step must be at least 4 feet.

It is important to understand that regardless of the maximum vertical allowance of a trench wall, the material must set the precedence for the digging to be done.

2.2.4 Taking Precautions

Accidents happen; they are inevitable. But many accidents are caused by poor preventive maintenance, neglect, or an operator not running the backhoe loader responsibly. Many accidents could be avoided if an operator would run the machine with one question in mind: "What would happen if?"
- An operator swings the boom straight at a laborer who is standing on the side of a trench. At the last second, the operator sends the bucket, full speed, into the trench. On one cycle the boom swings over at full speed, but as the operator hits the swing valve in the opposite direction to stop the boom, hydraulic oil explodes from the valve bank. The operator pushes the swing lever harder, trying to stop the boom, but it happens so fast that the operator doesn't have

time to think about slamming the boom to the ground to stop the momentum. Instead, the boom continues its arc and catches the laborer in the hips, throwing the worker 30 feet into a pile of trash. A return spring broke in the valve bank. The laborer suffers a broken pelvis and takes three months off to recover. The operator suffers a guilty conscience and a lawsuit.

- An operator lifts a worker out of a 4-foot hole with the bucket. The pressure line to the boom cylinder snaps and the bucket falls like a rock with the worker on it.
- An operator is loading trucks with the loader. There are a lot of trucks and a lot of material to load. To speed things up, the operator's full loader is high in the air when the operator decides to use the turning brakes to speed up the dump time. The rear tire spins on the rim, rolls off its bead, and goes flat at that very second. The backhoe rolls over and the operator is in trouble.

An operator's work is that person's trademark. It and it alone is what the operator needs to be concerned with when trying to demonstrate skills to a fellow worker or supervisor.

2.3   Traveling Public Roads

Driving a backhoe loader from point A to point B on a public road is called roading the machine. All operators will, at some point, find roading necessary or convenient. A backhoe loader is rarely licensed for travel on public roads, so roading is an occasional occurrence.

However few those trips may be, an operator should know how to travel safely on public roads. For example, the turning brakes should be locked. Keeping the brakes unlocked on the road invites disaster. For example, an operator may put a foot on just one brake, or perhaps the operator stepped in some oil before getting on the backhoe loader so that when he hits the brakes, his foot slips off.

Also, an operator should never allow traffic to build behind the backhoe loader, instead pulling over frequently to allow other vehicles to pass.

2.4   Changing Buckets

"Never operate this piece of equipment from any other place than the operator's seat." Most backhoe loaders have this warning somewhere in the operator's compartment with a caricature of an operator getting crushed by the boom.

Equipment manufacturers are right: Backhoe loaders should never be operated from the ground. But unless the operator has help, is extremely skilled, and none of the backhoe's rams leak, this warning is not practical when it is time to change buckets.

Chapter Two: Safety

When an operator reaches up to the controls from the ground, the operator is in a deadly pinch-point position. Adjusting the boom and crowd to change buckets from the ground places the operator in just such a position.

To reduce the possibility of an accident, the controls should never be operated from the ground at any r.p.m. other than idle, and the levers should be pulled slowly and individually. An operator should never pull two levers simultaneously to speed up the process. When pulling the lever, the operator should keep a free hand on the side of the boom. If the boom moves when the operator is pulling the crowd lever, the operator's hand will detect the mishap before it goes any farther.

If an operator finds it necessary to use a hammer to drive the pins in place, a piece of wood or other suitable material should be placed between the pin and the hammer. A metal-to-metal strike can, and eventually will, produce small shards of metal that are more than capable of imbedding in the operator's skin. An object between the two pieces of metal will eliminate this possibility of injury, as well as prevent the pin from mushrooming at the ends.

2.5 Safety Belt

Backhoe loaders sold in the United States must come equipped with a safety belt or harness. All backhoe loader manufacturers stress the use of the safety belt. The belt serves two important functions. First, it keeps the operator in a secure position enabling proper operation. Second, it helps support the operator's back when sitting in the seat.

A safety belt can save an operator's life by keeping the operator inside the cab if the machine rolls over or if there is a serious accident. It can also save an operator from injury when the backhoe loader travels too fast over a multiple set of railroad tracks or when it jumps a trench.

At the end of a day, an operator's seating position is often much different then the morning position. An operator's shoulders are a little lower, and there often is an air wedge between the crevice of the seat and the operator's lower back.

If the safety belt is tight, the operator's back will be in a supported vertical position, flat against the back of the seat. When properly supported, an operator should be less fatigued than if sitting in the slouched position all day.

2.6 Eye Protection

Everything a backhoe loader does creates dust: digging dirt, loading trucks with dirt or concrete, traveling down a road with a loader bucket full of sand. Operators need to wear eye protection to keep these materials out of their eyes.

24

Chapter Three:
Grade

3.1 Communication in the Field

Solid communication between an operator running a backhoe loader and a laborer on the ground is crucial to a successful job.

For example, an operator is cutting sidewalk grade and a laborer is checking grade on the ground. The operator takes a pass, sees the laborer's lips move, and nods back in acknowledgment. The operator makes another pass, the laborer's lips move, and again the operator nods back. The operator makes one more pass, but this time the laborer uses tools to check the grade. The laborer's lips move and points two fingers straight down. Again the operator nods and then takes one more pass. The laborer checks the grade again and nods, and the operator pushes back for a new pass.

Communication is the key to a good relationship between the operator and the laborer.

What makes this conversation unique is that the backhoe is operating at 2,500 r.p.m. and nobody within 20 feet of the machine can hear themselves think. But it appears that the operator can still hear because the operator seems to be doing exactly what the laborer is instructing.

But regardless of what the laborer thinks, the only instruction the operator took was the one for cutting the grade .2, and that was on the last pass. The previous conversations were completely one sided on the laborer's part. The operator had no idea what the laborer was saying until the laborer pointed two fingers straight down, giving the hand signal for a .2 cut in the grade.

3.1.1 Hand Signals

A backhoe loader operator must have a workable method for communicating with the laborer at all times and under all circumstances, and hand signals are the most effective form of communication. Although

Chapter Three: Grade

individual operators have their own set of hand signals they feel comfortable with, it is necessary to adapt the signals to each job and to make sure the grade checker (usually the laborer) is aware of them.

> *Converting tenths to inches is a matter of multiplying the tenths by 12. The equivalent of .30 is 36, or 3.6 inches.*

Basic hand signals work on the tenths system, using all eight fingers, both thumbs, a foot, and sometimes the waist.

All **fill signals** are made with the fingers pointing up; all **cut signals** are made with the fingers pointing down. The baby finger is used to indicate half tenths, while all other fingers are used to indicate full tenths.

- An index finger pointed up indicates a fill of .10.
- An index finger and a baby finger pointed up indicates a fill of .15.
- The first two fingers pointed up indicates a fill of .20. The first two fingers plus the baby finger pointed up indicates a fill of .25.
- The first two fingers and the thumb are used to indicate a fill of .30; the baby finger is added for an additional .05.
- Any four fingers on a hand are used for a fill of .40.

If, by this method, the baby finger on that same hand was used to attempt to indicate a fill of .45, the laborer would be sending an ambiguous signal with all five fingers pointed up, indicating a fill of .50, or half a foot. To avoid this confusion, a fill of .45 is indicated by four fingers on one hand and the baby finger on the opposite hand pointed up.

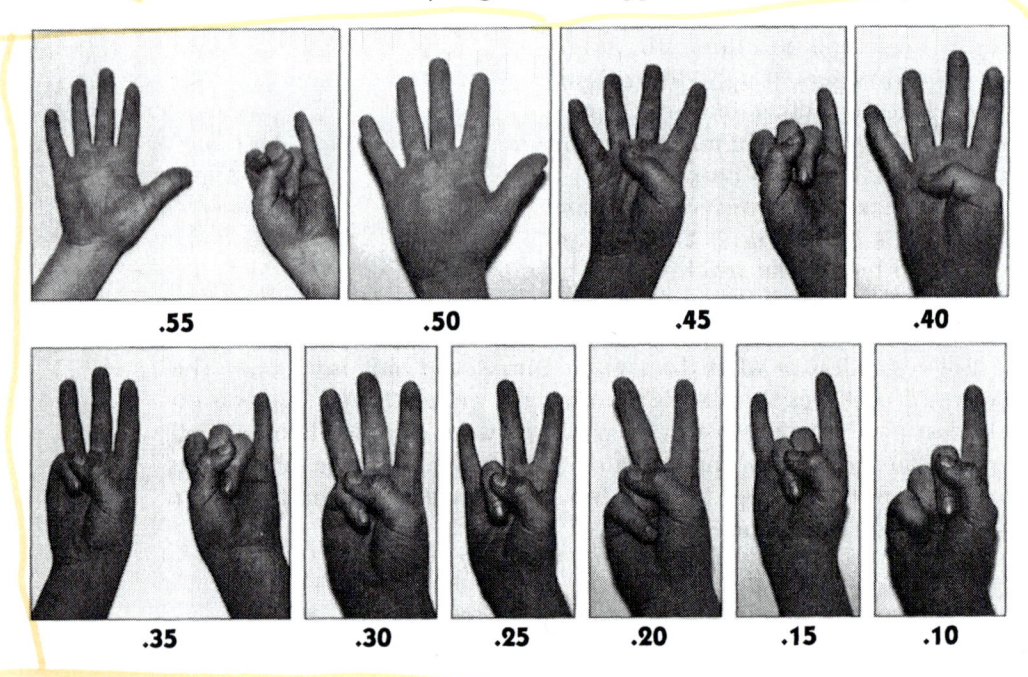

.55	.50	.45	.40
.35	.30	.25	.20	.15	.10

Basic hand signals.

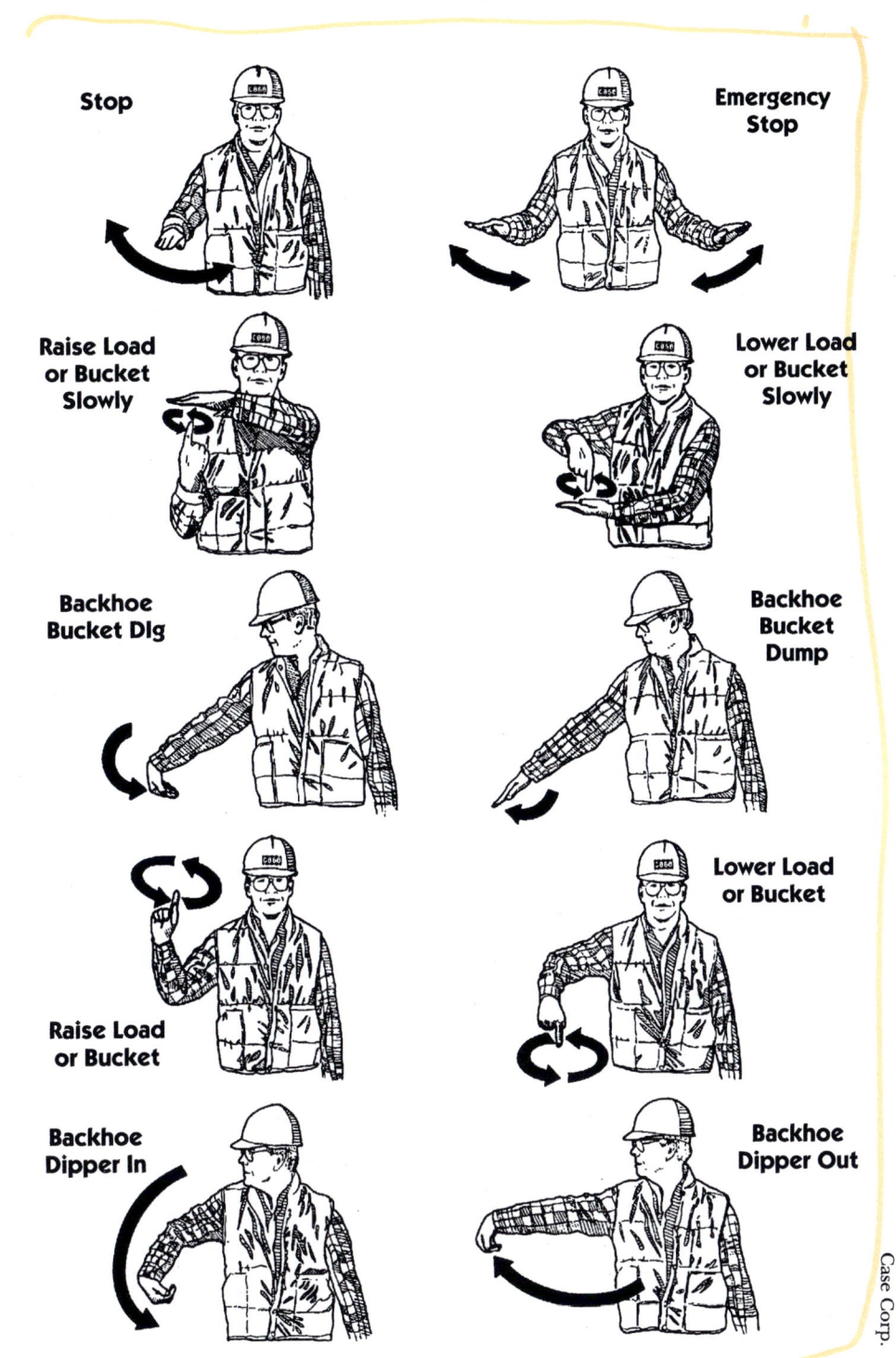

Communication in the Field

Stop

Emergency Stop

Raise Load or Bucket Slowly

Lower Load or Bucket Slowly

Backhoe Bucket Dig

Backhoe Bucket Dump

Raise Load or Bucket

Lower Load or Bucket

Backhoe Dipper In

Backhoe Dipper Out

Case Corp.

More basic hand signals.

Chapter Three: Grade

Turn Machine Left (Swing Load Left)

Turn Machine Right (Swing Load Right)

Raise Backhoe Boom

Lower Backhoe Boom

Come to Me

Move Away from Me

Start Engine

Stop Engine

Go This Far

All Stop and Hold

More basic hand signals.

For a fill of .50, the laborer uses all five fingers on one hand; the addition of the baby finger on the opposite hand would indicate a fill of .55.

For a fill of .60, .70, and .80, the laborer continues in the same way until the signal for .95 is needed. When a laborer needs to signal .95, a signal of either .90 or 1.0, a foot, is given. An operator will only cut or fill about .60 at a time, allowing the hand signal to be more precise when the material is adjusted a little closer to the grade.

A half of a foot, .50, can be indicated by the laborer cutting his or her foot or waist in half with a hand. A full foot can be indicated by the laborer bringing the foot up off the ground and pointing to it with a finger.

When the grade is very close and just needs to be cleaned, the laborer may make a brushing motion with a hand, indicating that the grade needs to be brushed up.

When an operator works with a crew or grade checker for the first time, they should review their hand signals to make certain they understand each other.

3.2 Reading Grade Stakes

To make the process of checking the grade quicker and easier, the tenths system is used. Unlike the American standard system where one foot is broken down into 12 equal inches, and the inches are broken down into .8 or .16, this system breaks the foot down into 10 equal parts called tenths. Each of the tenths are broken down equally into 10 units called hundredths. This system is used because it is much less complicated than a system of .8 and .16.

The front and back of a grade lath has numbers and abbreviations on it. It is important to understand what these abbreviations stand for, what

An operator should work carefully around grade stakes.

Chapter Three: Grade

the numbers signify, and how to apply them to specific grading jobs.

On the front of the lath at the top is a number with a line under it. This number is the offset, or distance, from the hub at the base of the lath to the side or front of the material being graded. Offsets are provided to assure a correct grade from a distance while the work is in progress.

Below the offset number is an abbreviation indicating the specific item being graded. [A few abbreviations are listed at the end of this chapter.]

Written lengthwise on the front of the lath is another set of numbers beginning with either the letter *C* or *F*. *C* indicates that the following number calls for a cut in the grade; the letter *F* indicates a fill.

The number following the *C* or the *F* is the specific amount the grade needs to be adjusted in order to start the project. This number has two parts: the larger number indicates feet and the two smaller numbers indicate tenths and hundredths.

For example, if the grade lath reads "C-1-23," then the grade needs to be adjusted to a point where the specific item being worked on is 1 foot, .20, and .03 below the hub at the base of the grade stake.

> *Operators should be careful around grade stakes because they are often the only aid an operator will have.*

On the back of the lath there are two sets of numbers divided by an addition sign. This set of numbers indicates the specific location of any given area or lath. The first number in the set is larger than the second set and gives the location of a section in hundreds of feet. It is used to find an approximate location in a specific area.

The second number in the set, indicated in feet, gives a specific point on a project. For long stretches of grade such as curb and gutter or sidewalk, the distance between the laths in the area being surveyed is generally equal in distance.

Station numbers are used as reference points, allowing the operator to find any particular point on a project by simply looking at the plans and then corresponding that point with the closest station mark. These numbers are important in case a series of laths gets knocked over and there is a question as to which lath belongs to which hub.

Sometimes a list containing all the cuts and fills of a section as they correspond to the station marks are provided with the grade plan. This is called a cut sheet. It is invaluable when large quantities of laths are knocked over or lost.

For example, in moving a pile of material, the operator accidentally knocks over a row of laths and moves them so that there is some question as to their original location. In order to complete the job, the operator needs the correct grades. After picking up the laths, the operator walks to the first standing lath in that row and looks at the station number. It reads "1845+25," indicating that the grade point relative to that project is 1,845 and the location of that lath is 25. But 25 what? The operator walks back one lath farther and looks at the station mark. The lath reads "1845+00." The 25-foot increment between the two consecutive stakes tells the operator that the laths are spaced 25 feet apart. All the operator needs to do now is sort through the laths until the next 25-foot increment, "1845+50," is found. The lath after that would read "1845+75."

3.2.1 Curb-and-gutter Grade

Banjo, hike-up stick, and cheater rod are all different names for the same tool used to help check grade. This tool maintains a constant vertical height and a level horizontal plane while the operator makes the necessary calculations to achieve the desired grade. Before any curb grade is cut, the operator should always take a look at the grade stakes to check the offsets and the cuts or fills. This will let the operator know if the dirt needs to be carried from a stake that has a cut on it to a following stake that has a fill on it. If the section is nothing but cuts, it will give the operator an idea of the amount of material that will be loaded, allowing the operator to make judgments regarding truck positioning.

> *An operator should not drop the banjo or leave it in the sun too long because the tool's accuracy will be affected.*

For example, the first stake reads 3, and below the 3, it says "face." This tells the operator that the hub at the bottom of the lath is 3 feet from the face of the curb. This is the offset.

In order for the operator to calculate the area that needs to be cut, it is necessary to know the type of curb-and-gutter grade being used. In this example, a standard 30-inch curb-and-gutter grade is being used. In addition to the 30 inches needed for the curb-and-gutter grade, the operator needs to add at least 6 inches on both sides of the curb-and-gutter grade for the concrete crew to set the form boards. Even if the curb is to be machine-poured, this extra width needs to be included because of the machine's sensors and to provide room for the back-finishing crews.

Starting back at the hub, the operator measures 3 feet and makes a mark on the ground. The lath shows that the offset is the distance to the face of the curb. A standard curb has a 6-inch top. The back of the curb is 2 feet, 6 inches from the hub. After the operator subtracts another 6 inches for the form boards, the inside cut is 2 feet from the hub.

To find the outer perimeter of the cut, the operator starts at the hub and adds 2 feet 6 inches to calculate the back of the curb. Another 30 inches is added for the curb-and-gutter grade, and another 6 inches is added for the form board for the gutter. The operator's additions make a 5 foot, 6 inch measurement from the hub to the outside of the cut.

Chapter Three: Grade

Once the curb-and-gutter grade area has been determined, the operator needs to figure the elevation. Below the offset and abbreviation are the words *cut* or *fill*. In this example, the lath reads "F0-50 T.C.," meaning a fill of .5 to the top of the curb. This means the operator needs to adjust the grade to a point where the top of the curb will be .5, or 6 inches, above the hub after the grade adjustment. The subgrade depends on three things: the height of the curb, the fill number on the lath, and the amount of hike-up on the banjo.

The operator uses a banjo without any hike-up on either end because of the short distance required to measure the curb-and-gutter grade.

The standard curb is 12 inches high at the back, and after the adjustment the top of the curb is 6 inches higher than the hub. This means that the subgrade needs to be 6 inches below the hub because the 6 inches below and 6 inches above the hub equals the 12-inch height of the curb. The operator makes the measurement at the back of the curb, which in this example is 2 feet, 6 inches away from the hub.

If the lath had read "C0-50 T.C." under the offset instead of the "F0-50 T.C.," it would have meant that the grade needed to be adjusted to a point where the top of the curb would be .5 or 6 inches below the hub. In this case, the operator would have taken the 12-inch thickness of the concrete, added 6 inches for the cut on the lath, equaling an overall cut of 18 inches from the hub to achieve the subgrade.

3.2.2 Sidewalk Grade

As a general rule, a sidewalk has a ¼ inch per foot hike-up to allow water drainage. This means that a 4-foot-wide sidewalk poured 4 inches thick has an outside edge 1 inch higher than where the sidewalk meets the back of the curb. The subgrade will match the finished surface of the sidewalk.

> *Operators should figure calculations carefully.*

To check grade, the operator can place a straight rod level across the top of the curb, extending to the outside edge of the grade cut. Directly behind the curb, from the bottom of the level rod, the tape measure should read 4 inches. Because of the ¼-inch-per-foot hike-up, the cut at the back of the sidewalk should be 3 inches.

What does the operator do when there is a 2-foot planter area between the back of the curb and the inside edge of the 4-foot sidewalk, and because of irrigation work, there is dirt spoiled over the top of the curb? The operator can shovel for an hour, clearing a flat spot from the top of the curb and through the spoils of dirt to the grade, or a banjo can be used. A good banjo will have an adjustable rod at one or both ends, allowing easy adjustment and calculations.

The operator should set the hike-up end of the banjo so that it is on top of the curb, set to facilitate calculations. The opposite end should be set at the desired grade so that all that the operator needs to do is set one end on the curb and the other end on the back of the grade. If the level that sits on top of the banjo is even, the grade is correct.

For example, an operator is checking grade for a sidewalk that is 4 feet wide and starts 2 feet from the back of the curb because of a planter. The operator needs to find the overall hike-up to the back of the sidewalk. Even though the sidewalk doesn't start at the back of the curb, the operator still includes the 2-foot planter in the hike-up calculation. At ¼-inch-per-foot, the operator adds the 2-foot planter area and the 4-foot sidewalk for an overall hike-up of 1½ inches.

Because of the piles in the planter area, the operator incorporates the banjo to reach over the piles to get the measurement. The banjo has a hike-up of 1 foot on the curb end and is set to the level position. If the banjo has an adjustable rod on the opposite end, it would be set at 14½ inches (including the thickness of the concrete). When the operator places the banjo on the ground at the back of the sidewalk, it should be level.

3.3 Sidewalk Grade

There are three things that need to stay in the forefront of an operator's thoughts when a sidewalk grade is being cut: a ¼-inch-per-foot hike-up, avoidance of water and gas lines, and careful outrigger and loader bucket placement.

If an operator pulls slightly away from the curb when grading, spoils will be kept to a minimum.

Chapter Three: Grade

The curb-and-gutter section is usually the first structure built on a construction site. For this reason, cutting sidewalk grade becomes much easier for the operator. The top of the curb can be used as a constant grade point from which the operator can derive one side of the sidewalk grade. On the downside, unprotected curb-and-gutter grade is extremely susceptible to damage from outriggers and loader buckets.

3.3.1 Hike-up

For every 1 foot of width, the sidewalk is elevated ¼-inch above the top of the curb. If the sidewalk is 4 feet wide, then the outside edge is 1 inch higher than the top of the curb measured at the back. The reason for this difference in elevation, or hike-up, is to allow water to drain off and not puddle, eliminating problems for pedestrian traffic.

> *An operator should pull slightly away from the curb when grading in order to keep spoils to a minimum.*

During grading, the ¼-inch-per-foot hike-up can easily be adjusted for by the outriggers. By raising or lowering either outrigger, the operator can adjust the angle of the backhoe which, in turn, transmits the change in angle to the teeth. With the help of the grade checker and a couple of practice passes, the operator can cut the desired hike-up in the grade without any further adjustment. This hike-up, adjusted for in the outriggers, can be maintained without further outrigger adjustment when the operator pushes back for successive passes.

3.3.2 Excess Material

Regardless of how close the material is to grade before the backhoe's teeth go through it, excess material will almost always be generated. Before any grading takes place, the operator needs to make arrangements for disposal of this material, whether in the street for a later pickup or in a dump truck as the area is graded. Any time a grade is cut there is potential for a great deal of shovel work because of the excess material being generated. Consequently, an operator needs help on the ground: Someone with a shovel to keep the dirt out of the gutter and off the top of the curb, to clean around meter boxes and other obstacles, and, obviously, to check the grade.

3.3.3 Setup

Ideally, the machine's setup should be slightly off-center toward the back of the future sidewalk. This off-center setup will force the extra material away from the curb and in toward the center of the grade as the bucket is pulled toward the operator on a pass. If an operator was to set up exactly parallel to the curb, as much material would spill over the top of the curb and into the gutter as would flow into the retrievable area of the grade.

Sidewalk Grade

Outriggers can chip the back of a curb.

A keyway is a continuous notch that runs along the back of a curb.

3.3.4 Outrigger Placement

Outriggers play an important role in the setup of a backhoe loader. They have the ability to chip the back of the curb and break the nose off the face of the curb. If the outriggers are rubberized, or if they have a protective underpadding, the operator can set them on the

Chapter Three: Grade

top of the curb without damage. This must be done carefully so that the weight is distributed equally between the back of the curb and the nose.

An exception to this is if the curb has a keyway running along the back of it. A keyway is a continuous notch poured into the back of the curb. This notch is designed to keep the sidewalk from lifting or separating from the curb should ground swelling occur under the sidewalk. If an outrigger is placed on a curb section that contains a keyway, it can break the back off of the curb.

The loader should remain elevated whenever cutting a grade that would put it in contact with the curb, keyway or not.

3.3.5 Grading

An operator's first sidewalk pass should be taken at least 1 foot from the back of the curb. By removing this dirt first, the dirt on the next pass, which will be right along the back of the curb, can flow away from the curb into a workable area instead of over the top of the curb and into the gutter. Only when this dirt has been removed from the center area should the operator start making grade passes along the back of the curb.

> At the wrong angle, the back of the shank can chip the back of the curb as the shank is brought in. The bucket must be in the full position when a pass is made along the back of the curb. If the shank is still rubbing, the backhoe should be offset at a greater angle relative to the curb.

With the backhoe set at the off-center position toward the back of the sidewalk grade, the operator should make the first pass along the back of the curb. When the grade checker indicates that the desired grade has been reached along the back of the curb, the operator should open the bucket fully and gently lay the outside tooth against the back of the curb, with the tips of the teeth just touching the finished sidewalk grade. The operator needs to take note where the outside tooth lies against the back of the curb and make an imaginary mark on the tooth at this point. This imaginary mark will be the point from which the grade is derived along the back of the curb.

After the grade pass has been made along the back of the curb, the operator should put all loose material, such as a windrow, into a pile and dispose of it.

On the next pass, the inside edge of the bucket, the edge closest to the curb, should be moved over about a foot from the outside edge of the last pass. This pass overlaps the last pass by about a foot.

The third pass should be made against the outside edge of the sidewalk area, an extra 6 inches cut for the sidewalk form board.

The final pass for the sidewalk section should be a cleanup pass running down the center of the grade to pick up any loose material and drag it into the next grade section. During this pass, the previous grade passes on the inside and outside edges of the sidewalk area should be the grade on which the operator rides the teeth, as the center section of the bucket cuts the remaining grade.

Assuming it requires an average cut, there should be three or four passes per setup when grading sidewalk for a production concrete crew.

In order to maintain the hike-up established by the adjustment of the outriggers, when the operator prepares to push back, the operator should boom straight down to elevate the machine instead of raising the outriggers. This eliminates the need to check and adjust the angle of the outriggers when setting up for the next pass. It also saves time. The operator needs to make sure one of the outriggers is not pressed up against a meter box or some other obstacle that could sustain damage.

3.3.6 Obstacles

Obstacles make cutting sidewalk grade challenging. Operators need to be aware of obstacles in back and in front of the backhoe loader. Even when operators think they know the area well, they need to look twice and watch in the direction of movement when pushing the backhoe loader for a new pass.

Location of obstacles, such as meter boxes, and the width of the bucket play a role in determining how to grade the area. If the distance between the curb and the meter box allows the bucket to pass through unobstructed, the operator need not worry. If the distance between the obstacle and the curb is too narrow for the bucket to pass through, the operator can somewhat offset the angle of the backhoe, thereby narrowing the profile of the bucket relative to the grade. The greater the angle of the backhoe relative to the grade, the narrower the width of the pass made by the bucket.

> *The greater the angle of the backhoe is relative to the curb, the narrower the width of the pass.*

When a light or telephone pole is in the grade, the operator should grade up to it as close as possible. After the operator cannot dig any farther, the operator should pull off of the grade and approach it with the loader at a perpendicular angle.

3.3.7 Using the Loader

The operator should line up the edge of the loader with the point left off with the backhoe bucket, slowly easing the cutting edge of the loader into the grade directly behind the curb at a 90-degree angle. The loader bucket should sink into the grade until the cutting edge matches the grade made with the backhoe bucket. If the cutting edge has difficulty sinking down far enough to match the grade, the operator should apply more down pressure to the loader and slightly wiggle the loader in a back-and-forth motion. This should push the loader farther into the grade. If the extra movement of the loader still does

Chapter Three: Grade

not help, the operator will need to cut that section of the grade.

After the existing grade has been matched with the loader, the operator should stop and tap back on the loader control handle until it starts to raise from the grade, slowly returning the loader to its position with the cutting edge matching the existing grade. This needs to be done, because as the operator applies pressure to force the cutting edge deeper into the grade, hydraulic fluid is being forced from the loader cylinder rams into the hydraulic reservoir. When the loader reaches the desired depth, the operator should release the loader control handle. This closes the valves leading to the loader cylinder rams and locks the remaining fluid in the rams under pressure. If the operator was to drive forward with the cutting edge at the matched depth, the pressurized fluid locked in the cylinders would cause the loader to cut deeper into the grade. Tapping back on the loader control handle releases the pressurized fluid from the rams, allowing instant control of the loader.

The loader pins should not contact the back of the curb when the operator curls the loader bucket to the 3 o'clock position.

The operator should then slowly drive the backhoe forward a foot or so, rolling the loader to the 3 o'clock position, adjusting the elevation of the loader if necessary to remain at the same matched grade, and then slowly drive forward another foot or so again.

After rolling the loader bucket to the 3 o'clock position and driving forward, allowing the material to fill the loader bucket, the operator should continue driving forward until the front cutting edge is well outside the sidewalk area. In order to push the loader bucket far enough toward the back of the grade and not interfere with the loader grade, the operator may need to push the loader control lever all the way forward to elevate the front wheels high enough to clear the top of the curb. The material in the loader bucket can then be thrown away, allowing the operator to set up on the other side of the obstacle to continue grading the sidewalk.

3.4 Curb-and-gutter Grade

The techniques for cutting curb-and-gutter grade and cutting sidewalk grade are very similar. The only major difference between the two is that in cutting sidewalk grade the operator normally has an existing structure to grade by, such as the back of the curb. A curb-and-gutter does not.

With curb-and-gutter grade, the operator has the grade stakes or grade laths to grade by. Having a working knowledge of grade checking can come in handy during the grading process.

Depending where the curb-and-gutter grade is to be cut, the operator needs to survey the area for any obstacles or potential problems,

An operator should pay close attention to valves or meter boxes when grading.

keeping in mind that every house has water and gas lines. Normally, both of these utilities are low enough to avoid teeth contact, but at the point where they raise from the main line to enter a box or meter, they could be shallow enough to catch the backhoe's teeth.

3.4.1 Setup

A look at the area to be graded will tell the operator how to set up for the job.

If the work will be done in a new construction site where there are no existing sidewalks or asphalt streets, the operator can set up straight on, straddling the area to be graded with the loader down for good stability.

If the grade section is in an area next to an existing sidewalk or street section, the operator will need to make a judgment about the setup, taking certain variables into account.

Outriggers and the loader bucket are two important variables when setting up next to existing structures. Setting up straight on the grade with the backhoe in a straddling position over the grade puts the outriggers and the loader in an ideal position to crack or chip the sidewalk when weight is placed upon them.

If the operator is not careful, utility boxes can fall victim to the outriggers, loader, or tires when cutting grade straight on. The exception is if the sidewalk or utility boxes are at an offset to the grade.

3.4.2 Hike-down

Looking at a piece of curb-and-gutter grade at a cross section shows that the point directly below the curb face, the flowline, is actually low-

Chapter Three: Grade

er than the lip of the gutter. This angling back of the gutter is called the hike-down, and it is what creates the water flowline.

Operators are sometimes required to cut the curb-and-gutter grade at this hike-down angle. With the help of a grade checker, this angle can be adjusted for in the outriggers. (This is similar to adjusting for the hike-up when cutting sidewalk grade.) Under ideal conditions, once the hike-down in grade has been established through the outriggers, the operator should be able to grade the area, push back, and continue grading without readjusting the outriggers.

Ideal conditions are rare, so operators need to be prepared for the time when, because of an obstruction in the grade or because the backhoe may need to be offset at a different angle, the angle of the outriggers needs to be changed to compensate for the grade.

The operator will also need to remove excess grade material. Either the material will have to be put directly into a dump truck, or it will need to be put off to the side for later removal.

Normally, in a removal and replacement (R&R) or a revamp (updating an existing structure), a rough subgrade has already been cut, leaving either a minimal cut on the bottom or a minimal cut in the back. In either case, not much material is being moved. The operator needs to plan on grading and pushing the backhoe ahead while pulling the material with the backhoe until enough material has accumulated to warrant taking the time to put a couple of bucket fulls into the truck or off to the side. If the material is being put in the street to be picked up later, the operator should keep it far enough away from the grade so that the dirt is not pushed into the new grade when it is being picked up.

3.4.3 Grading

There are many different types of curb, each one tailored to a specific purpose. In order to grade the section in preparation for the curb, the operator needs to know the type being used.

For example, a standard type C curb-and-gutter grade has a 12-inch back, a 6-inch top, and an 8-inch face. The gutter is 24 inches long, and the curb-and-gutter's overall width is 30 inches.

In this example, the project is in a new residential development and the street subgrade has been cut. The operator (on the motor grader) that made the subgrade extended the cut on both sides of the street by 3 feet and removed the heavy material for the curb-and-gutter grade, creating a curb bench.

The surveyors have set the grade stakes, and the operator takes a quick look at a few of the grade laths to check the offset and determine if the majority of the laths read cuts or fills.

The grade laths have a 3-foot offset, and most read cuts, with a couple of fills. This tells the operator that material will be removed, not

The curb bench pictured above needs a little adjustment to achieve grade.

imported.

The placement of the grade laths and the position of the rear tires against the curb bench prohibits the operator from setting up straight on. Instead, an off-center position with the backhoe at about a 20-degree angle is the best pose. Because it is a new construction site and the loader bucket is in the street section, the operator's only concern is the placement of the outriggers; they need to be kept away from the grade laths.

The operator knows some constants will not change. First, the offset, the distance from the hub at the base of the lath to the face of the curb, is 3 feet. But the top of the curb is 6 inches wide, putting the back of the curb 2 feet, 6 inches away from the hub. This is the true offset distance.

The cut distance is calculated by starting at the offset distance of 2 feet, 6 inches, adding 30 inches for the overall width of the curb, and adding form board room. Form board room, which is for the concrete crews, requires the operator to subtract 6 inches from the back of the curb and add 6 inches to the gutter side of the curb, equaling a starting cut of 2 feet from the hub and an ending cut of 5 feet, 6 inches from the hub.

Another constant is the height of the curb which, in this example, is 12 inches measured at the back.

The cut or fill amount on the lath is the distance from the hub to the top of the curb. Because the operator is cutting subgrade, 12 inches must be added to all calculations.

Chapter Three: Grade

There is only one variable: The amount the grade will need to be adjusted in order to correspond with the cut or fill numbers on the grade lath.

The first grade lath reads <u>3</u>. This is the offset. Directly under that are the initials *C.F.*, indicating that the offset distance is measured to the curb face. Written lengthwise below that is *C-1.22*. The grade needs to be adjusted to a point where the top of the curb is 1 foot, .22 below the hub. Adding in the height of the curb, the subgrade is 2.22 below the hub. The present grade is 2 feet below the hub, only .22 from subgrade.

The first cuts are to trim both sides of the grade to accommodate the entire width of the curb-and-gutter grade, including the 6 inches on either side for the form boards. The operator should start at the lath side first and take that side down to the existing grade, 2 feet away from the hub. The next pass should be on the far side of the section, with the backhoe's outside tooth cutting 5 feet, 6 inches away from the hub. This side should be brought down to existing grade as well.

The next pass should be back over to the lath side, a single .22 pass. This pass should be copied on the far side, and a cleanup pass should be taken in the middle. The measure memo should read 2.22.

Since the operator is set up in the off-center position, the backhoe loader cannot be pushed straight back. The operator needs to set the backhoe bucket next to the section just graded, about as far over as the outside edge of the loader bucket. Making sure the loader bucket is set firmly on the ground and boom down, the operator should lift the machine just high enough to clear the grade. The backhoe should then be swung over and away from the grade until it is elevated in a straight line with the loader.

At this point the backhoe should be hovering parallel to the grade. After slowly lifting the loader and pushing back, the loader should be firmly on the ground and pushed to the correct angle.

3.5 Grade Stake Abbreviations

AC	Asbestosized concrete
Adj	Adjust
BC	Begin horizontal curve
BCR	Begin curb return
Beg	Begin
Bk	Back
Bldg	Building
BM	Bench mark
C&G	Curb-and-gutter
CL	Centerline
Culv	Culvert
Dr	Driveway

Grade Stake Abbreviations

Ease	Easement
EC	End horizontal curve
ECR	End curb return
Elev	Elevation
EP	Edge of pavement
FG	Finish grade
FL	Flowline
Ftg	Footing
GM	Guide marker
Irr	Irrigation
Loc	Location
Ln	Line
PB	Pullbox
PL	Property line
POC	Point on horizontal curve
POT	Point on tangent
POVC	Point on vertical curve
R	Radius
RP	Reference point
RW	Retaining wall
R/W	Right of way
SL	Station line
Sec	Section
SG	Subgrade
Sta	Station
Str	Structure
SW	Sidewalk
Swr	Sewer
TG	Top of grade
Trans	Transition
TS	Traffic signal
VC	Vertical curve
Xing	Crossing

Chapter Four:
Dirt Work

4.1 Trenching

This chapter concentrates on shortcuts to help the operator refine the backhoe movements to increase speed and achieve greater overall efficiency.

> *An operator's goal is smoothness. Speed will follow.*

4.1.1 Cycling

A cycle is defined as the time from when the backhoe's bucket teeth sink into the ground to the time the bucket returns to its starting point, having emptied the material. This cycle should be done in a large, circular motion. The faster the operator can get the backhoe to cycle, the faster the job can be done.

More important than speed, however, is the smoothness of all backhoe operations.

Movements should resemble an arm and hand gracefully lifting buckets of material, and not the bucket slamming around because the operator is digging as fast as possible. Speed can be gained by addressing the individual actions of a cycle and then eliminating the unnecessary movements.

> *Unless the trench is unusually deep, the grade checker should take the grade measurements from the top. This will leave the operator with a clean line of sight when pushing back to continue.*

The individual movements of a backhoe in the process of digging must be understood before improvements can be made to the motions of a cycle.
- The boom and crowd arm or dipper stick are extended just short of the overall reach.
- The bucket is opened or extended and the boom is lowered onto the grade.
- Pressure is applied to the boom while the crowd arm and boom are pulled in simultaneously, drawing a straight grade in the material.
- The bucket is curled, holding the material from the trench.

Chapter Four: Dirt Work

- The boom is raised high enough to clear the trench wall.
- The boom is swung past the side of the trench.
- The bucket is opened and the material is released.
- The bucket is returned to the trench for the next pass.

Speed and efficiency can be gained by incorporating three rules of trenching. First, the shortest distance between two points is a straight line. Second, the more valves that can be incorporated into an action, the smoother the digging will be. Third, as many tasks as possible need to be accomplished in as few moves as possible.

By incorporating the following rules, the movements listed above change accordingly.

- The boom and crowd arm is extended just short of the overall reach.
- The bucket is opened or extended so that the teeth are in a straight up and down position as the boom is lowered onto the grade.
- The crowd arm is pulled in slightly to start the teeth into the grade while the boom compensates for the crowd arm's desire to gouge the grade.
- The teeth are kept in the vertical position for a couple of feet until a constant depth has been established for the trench.

4.1.2 Digging Sequence

Once this is done, the operator can incorporate the bucket into the digging sequence. The bucket should start to curl 2 or 3 feet into the pass, and it should be timed so that the bucket is curled enough to hold the material about 3 feet short of the innermost digging point.

As the trenching progresses, the operator should keep the material from building up too high between the outriggers, also making it a habit to pull the bucket back slightly away from the inside end of the trench before fully curling the bucket to swing out of the trench. Before bringing the bucket out of the trench, the operator needs to mentally draw a straight line from the bucket in the trench to the point where the material will be released, having the bucket follow that line as it is pulled out of the trench. This means the operator needs to boom up and swing at the same time. When done correctly, the bottom of the bucket will almost brush the top edge of the trench as the bucket swings out.

This brushing will come in handy as the trench becomes deeper. The side of the bucket will push the material located close to the side of the trench back with the rest of the spoil. This is a good example of accomplishing two goals with one movement.

The bucket will start to open or extend the instant it passes the side of the trench. After swinging over the spoils and bringing the bucket to the proper position for the next pass, the operator should be sure the teeth are in the vertical position.

The material will fall out as the cycle continues so there is no need for the operator to open or extend the bucket all the way to release the material. Depending on the bucket's profile, it generally does not need to be opened more than about halfway. If the material is sticky or claylike, the crowd arm can be used to dislodge the material.

As the bucket is passing the side of the trench and is starting to open, the crowd arm should be extended so that the material can fall from the bucket onto the spoil bank. The operator should continue to extend the crowd arm out to the starting point of the pass, and the boom should be swung over. The bucket should re-enter the trench for the next pass, with the teeth just above the side of the trench.

In a good, smooth trenching cycle, there are three or four valves being worked simultaneously. It is difficult and tiresome on an operator's hands, but in the long run the work will pay off because when the operator can use the backhoe smoothly, the work will get done faster.

4.1.3 Confinement

There are good reasons to confine trenching to a small area. First, as a backhoe gets more use, the bushings wear and the pins begin to get worn and sloppy. This occurs in every bushing and pin from the swing cylinders to the bucket wrist pins. This translates into a great deal of play when the crowd arm and boom are fully extended.

Second, a scalloped trench (a trench that curves at each setup) will result when an operator reaches out as far as possible. Digging in as close as possible, or digging without the backhoe in level posture, will produce the same results.

Third, if the operator has pushed back for perspective and discovers a correction needs to be made, the operator can do so without having to pull to reach where it is needed to start that correction.

Once the operator has veered off the chalk line, the machine needs to be straightened out. The operator should not reach back and correct only the area that needs to be corrected. The backhoe should be extended another 5 feet to make a smooth transition into the area that needs to be corrected, and an additional 5 feet to move back to the original chalk line. This will make the trench look better than having a jog out where the error was corrected.

4.2 Straight-line Trenching

There may be times when the straightness of a trench isn't a concern to others on the jobsite. They just want the operator to trench from point A to point B and not bother with chalk or string line. Even if the distance from A to B is 200 feet, with 1-inch conduit being placed in an 18-inch trench, the operator must still be straight.

The best solution is for the operator to drive one stake at point A, the starting point, another at point B, the ending point, and the third

Chapter Four: Dirt Work

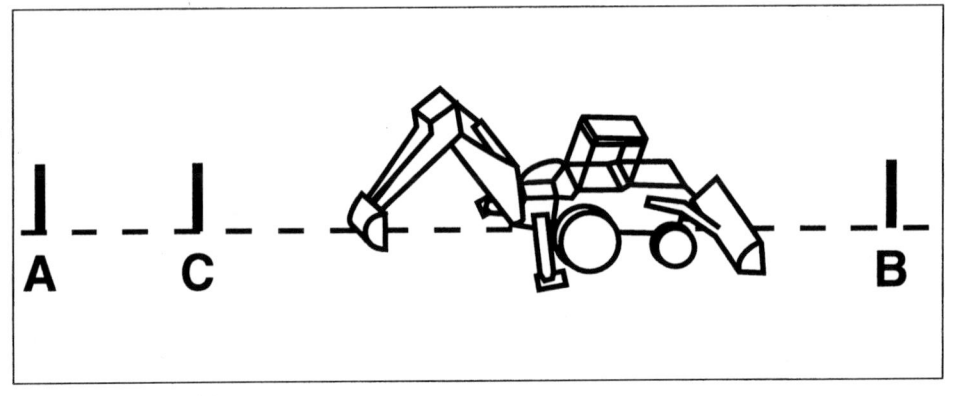

Straight line trenching.

at C, about 20 feet behind point A. Stake C should be lined up with stakes A and B, hiding them.

The operator should set the backhoe loader in a normal trenching position and dig with the backhoe boom aligned on stakes A and C, with the center of the loader aligned with stake B.

4.3 Pushing Back

Folding up a backhoe and turning it around every time the operator needs to advance during trenching would be very time-consuming. A backhoe loader's hydraulic system allows the operator to continue digging while pushing back.

First, before trenching begins, the operator should be aware of all utilities in the area. This is essential not only to avoid rupturing underground lines, but also to eliminate the possibility of puncturing lines with bucket teeth if the operator should decide to use the bottom of the trench as the push-off point.

> Whenever possible, the operator should use the top edge of the trench to push back on. Using the bottom of the footing/trench creates an extra pass for the bucket to reach back and smooth the gouge marks.

For example, a trench is 18 inches deep and 12 inches wide. After the operator has made the grade, the backhoe loader needs to be pushed back. The operator should first look for utilities to prevent rupturing an underground line or puncturing a line with the teeth. Now the operator must make a decision: Should the bucket be placed in the trench to help push the machine back or should the top side of the trench be used to push back?

The advantage to using the bottom of the trench as the push point is that it is quicker because the circular digging pattern is not being interrupted. After the last cleanup pass, the operator can bring the bucket into the trench as for another pass. Instead of extending out, the bucket is brought to the halfway point, opened, and the teeth sunk into the grade.

Backhoes loaders are heavy; it takes quite a bit of force to get one rolling. This same force is transmitted to the teeth. If there is anything underground within 1 to 2 feet of the teeth, they will go through it.

The material which is being dug in and the depth of the trench also need to be taken into consideration. If the backhoe is digging in a heavy clay or adobe soil, there is not much to worry about. In sandy or lighter loam soils, more care must be taken when deciding where to place the bucket. For lighter material, the bucket needs to be set far enough away from the edge of the trench so that it will not cave in. But if the machine is set too far away, the swing cylinders will be incorporated when the operator pushes back, and the operator will end up fighting to keep the backhoe loader straight on the chalk line.

In general, it is faster and more efficient to use the bottom of the trench as the push point. The exception to this is if there is a utility in the area or if the trench is excessively deep.

If speed is an issue, the operator can trench with the loader bucket elevated a couple of feet in the air and the parking brake released. When it comes time to push back, the operator has to make sure the front tires are straight, the outriggers are slightly elevated, and a soft push is given using the backhoe bucket to stop when the desired distance is reached.

4.4 Digging Under Utilities

Once the operator has located the utility and is ready to continue the trench by going under it, a clean field of view to the conduit is needed and must be maintained until all digging around the conduit has been completed.

Digging under a utility requires removing the material from each side of the conduit and from underneath it. This must be completed in a manner that allows the operator to maintain visual contact with the utility at all times. It is easier and safer to dig out one side of the conduit and then set up the backhoe loader on the other side and dig it out. Setting up twice to dig around one conduit is not efficient and should be considered a last resort.

> **SAFETY TIP**
> Underground conduits are often hidden near curb-and-gutter grade.

4.4.1 Starting the Dig

The backhoe loader should be situated so that the conduit is at one-half the distance of the backhoe's overall reach. The first pass should start as close to the conduit as possible, with the bucket on the side closest to the operator. The bucket should be opened or extended so

Chapter Four: Dirt Work

that the teeth are at the farthest point back. Keeping the bucket in the full open position, the operator should pull the material directly toward the backhoe, a few inches at a time. Immediately curling the bucket in order to remove material too close to the conduit risks damaging the conduit with the back of the bucket.

Ideally, the operator should dig between 12 and 18 inches deep in front of the conduit to allow the operator to punch the teeth under the conduit at a safe distance. Although this depth may not be possible, it is ideal because the deeper the operator can dig, the less likely the conduit will be damaged.

An operator should always keep the utility in sight.

Digging Under Utilities

With the material on the front side of the conduit removed, the operator should carefully place the tips of the teeth, with the bucket in the half open position, on the far side of the conduit and boom straight down. The teeth should be sunk just a couple of inches into the material and crowded out, pushing the material straight back.

The goal of this stroke is to keep the material from piling up and covering the conduit when the operator takes the first stroke forward, in back of the conduit. The operator should start a couple of feet behind the conduit and sink the teeth straight down into the grade to prevent buildup.

Curling the bucket until the material starts to raise around the conduit, the operator should crowd out the same amount as was the bucket curled up. When the bucket was curled up, the back of the bucket should have left an imprint on the rear wall of the material behind the bucket. The operator should crowd out until the bucket makes contact with the imprint on the rear wall. The operator should then boom straight up, curling the bucket only after the tips of the teeth have cleared the conduit. After the bucket of dirt has been dumped, the teeth should be placed directly behind the conduit, with the bucket in the half open position.

The operator should boom straight down 8 to 10 inches and crowd out, pushing the material into the receiving hole just created. The operator should continue these two steps until the depth behind the conduit matches or is slightly deeper than the grade in front of the conduit.

After the bucket has been lowered into the hole and placed in a position where the top edge of its sidewall is level, the tips of the teeth should be brought just under the conduit and pulled in toward the operator, allowing the outside front floor of the bucket to push the material from under the conduit. Boom pressure may need to be used if the bucket seems like it wants to ride up on material under the conduit instead of pushing it. It is crucial that the operator makes sure the top edges of the bucket walls do not come in contact with the conduit while the bucket is being pulled underneath the conduit. If the top edges of the bucket rub along the bottom of the conduit, the conduit can easily be broken by the friction produced by the rubbing or from the pressure of the sides themselves.

Once the operator has pulled in as far as the conduit will allow, the steps should be reversed and the bucket allowed to exit from underneath the conduit.

By this point most of the dirt should have been removed from under the conduit. If there is any dirt remaining, the operator should bring the bucket back in directly under the conduit. This time, instead of keeping the bucket level, it should be opened until the bucket floor is parallel with the trench grade. The operator should boom straight down until the tips of the teeth make contact with the bottom of the

Chapter Four: Dirt Work

trench and crowd straight back, letting the teeth push the material out from underneath the conduit. (When crowding straight back from under the conduit, the operator may need to raise the boom slightly for compensation if the back of the bucket starts to dig in.) Any material left in front of the conduit can be reached from a forward position when the backhoe loader is pushed back to continue the trench.

4.5 Digging Under Curb-and-gutter

The technique used to dig under a conduit can be expanded or modified to be applicable to a variety of digging needs. Digging under curb-and-gutter follows the same basic steps as digging under a conduit, with one exception: After the material has been removed from the front of the conduit and a relief hole has been dug in back of the conduit, the center material directly under the conduit should be removed by digging right under the conduit and working to the desired grade. When digging under curb-and-gutter, the operator should start at the bottom of the trench and work up to the bottom of the curb-and-gutter.

> *When digging under curb-and-gutter, the operator should watch the crowd collar against the gutter so that the curb does not crack.*

When digging a relief hole behind the curb, the deeper the operator can dig, the farther under the curb the bucket will be able to curl, enabling the operator to reach more material.

A word of caution: Curb-and-gutter can crack or break very easily, especially at the joints. To avoid these cracks and breaks, an operator needs to keep two important safeguards in mind. First, a section of undisturbed material should always be kept between the bottom of the curb-and-gutter and the area being dug. This section of undisturbed material acts as a cushion between the bucket and the curb-and-gutter. Second, all of the bucket power strokes need to be in a downward motion, pushing the material away from the curb-and-gutter and thereby eliminating the possibility of accidentally curling the bucket too far and cracking the curb.

> *If the bucket's top edges rub along the bottom of the conduit as the operator pulls in, the rubbing can cause the conduit to break.*

4.5.1 The Relief Hole

After digging a sufficient relief hole behind the curb, the operator should set the bucket in a ¾-open position and place the teeth at a minimum of 1 foot below the bottom of the curb. The operator should push the teeth into the material a few inches and open the bucket, letting the natural opening motion of the bucket push the material into the relief hole.

After repeating this step a couple of times, the bucket should be set flat with the bucket floor parallel to the bottom of the trench. Because the curb will probably be blocking the view of the bucket, an inexperienced operator should climb down from the backhoe loader, visually check the position of the bucket, and adjust the position accordingly. The operator should draw the bucket in, riding against the bottom of the grade and picking up the loose material.

> *An operator should dig under the curb at a score line whenever possible. If the curb cracks, it will do so at that point.*

The operator should continue to draw in until contact is made with the undisturbed material under the curb. After a change in pressure has been made from the contact, the operator should curl up slightly, breaking loose more material, and then back straight out of the relief hole and dump the bucket. The bucket should be brought back in again and placed in the ¾-open position. The tips of the teeth should be brought to a point a foot below the bottom of the curb in order to penetrate the material 2 or 3 inches. Opening the bucket, the operator should push the material down and away from the curb to the bottom of the trench. With the bucket back in the ¾-open position, the operator should lower it to the bottom of the trench, with the teeth just touching the grade.

After drawing in and penetrating the bottom material 2 or 3 inches, the operator should slightly curl the bucket up, breaking the material loose. The bucket should be backed up, lain flat, and the rest of the loose material retrieved.

The pattern taking shape is an easy one. It is simply a matter of working the teeth to loosen the material by using a downward stroke from the top and a slight upward movement from the bottom. Once a hole has been established, the operator should work from the top of the new hole, and with bucket in a position to where the sides are parallel with the bottom of the curb, work the material down toward the bottom of the trench until the hole under the curb is of desired size.

4.6 Footings

4.6.1 House Footings

Before any digging starts for house footings, the backhoe loader operator needs to formulate a plan of attack by taking a good look at the foundation lines about to be trenched. The operator will then need to decide where to begin, where to end, and at what point during the digging, if any, the interior work will be done.

> *Operators should be careful not to dig themselves into any corners.*

When they are available, an operator should use a set of plans; they can save a lot of leg work and may help in avoiding simple mistakes, such as overlooking an interior footing or forgetting to dig a fireplace jog out.

Chapter Four: Dirt Work

4.6.2 Line Location

The footings will usually be marked on the dirt with either chalk or paint. The operator may need to ask a supervisor whether the line indicates left side, right side, centerline of the trench, or any offsets. The supervisor may ask what line location the operator prefers prior to chalking out the foundation. If this happens, the operator should remember what side of the backhoe he or she naturally looks over when digging. This can save a lot of neck pain.

> *An operator should accomplish as many tasks as possible in as few moves as possible.*

4.6.3 Adverse Material

Square corners, smooth bottoms, and clean edges are a must when digging a foundation. The material in which the foundation is being dug should be examined before digging. If the material is excessively rocky or contains a large amount of dry sand, the operator should mention it to the foreman so that everyone is aware that the material could affect the shape and width of the footing.

In most cases the footing is part or all of the actual form for the poured concrete. More concrete will be needed to displace the dirt the wider or deeper the footing is dug. Concrete costs money, so it is best for the operator to offer this possibility before digging is started. Clean square corners not only save money by reducing the volume of concrete, they also highlight an operator's ability.

4.6.4 Formulating a Plan

While looking at the plans or formulating a digging approach by surveying the chalk lines, an operator should always be looking for ways to save time or for shortcuts. The goal is to try to accomplish as many things in as few moves as possible.

An operator should choose the line location that is easiest.

For example, operators should look for interior footings that run parallel with the foundation footing. If the interior footing is close enough, it may be easier and quicker to dig the interior footing at the same time as the exterior foundation footing.

When this is the case, the backhoe loader should be set up between the two footings so that it can be lifted and moved to one footing so that the operator can dig as much as the reach will allow. With the loader still planted firmly on the ground, the backhoe should be moved over to dig the second footing. In addition, the backhoe will need to be in a position to dispose the material from both footings without making an extra move.

> **SAFETY TIP**
> If a laborer is checking grade in the footing behind the bucket, the operator should be sure the laborer knows every time the operator moves the bucket back for a pass. This is especially important if the laborer is backed into a footing corner.

4.6.5 Executing the Plan

The starting point will probably be a corner of the foundation that can only be reached from one side because it is too close to any existing structure. The operator should dig one long side of the foundation footing first, followed by interior digging, the sides, and the final edge. This pattern should be used because the operator will probably need to cross the front foundation line to remove dirt from the interior footings or piers. When traveling in and out of the foundation line area, operators should try to cross the foundation chalk line at about the same place each time. This will prevent the operator from having to take time to rechalk the entire front foundation line.

Regardless of whether the operator is digging left side, right side, or centerline, the backhoe loader should be set up dead center on the footing line, with the line splitting the backhoe down the middle. If the line is being dug off-center, the loader should be placed firmly on the ground and the backhoe lifted high enough to prevent smearing of the chalk line with the tires or outriggers. The operator should then move to the side of the line being dug. It may be helpful to position the backhoe at a slight angle, giving the operator an easier line of sight to the chalk line around the boom.

When reaching to the corner for the first digging pass, the operator should start with the teeth just short of the corner, leaving an opposing line. The first pass should be a ghosting pass, a pass where the teeth ride just above the surface as the operator draws in with the bucket. This will allow the operator to see if the bucket is set up correctly and if the operator is drawing the bucket straight on the chalk line.

The first digging pass should be just shy of the opposing corner chalk line, a few inches deep and level. (The corner line can be taken

Chapter Four: Dirt Work

A ghosting pass can either ride above the surface or barely touch it.

on the last pass after the trench has been dug to its grade depth.) It is imperative that this first pass be straight because the next pass is determined in part by the walls formed on this pass. As the footing progresses, it can be very time-consuming to correct a crooked trench. When it is time for the next corner, the operator should not spend time trying to square it, because it can be cleaned up when trenching in the following foundation side.

When trenching in the first side and approaching corner, the operator should dig up to the bisecting line, coming as close as possible while still leaving the actual line. Leaving the line allows the operator a clean, straight line to start from when setting up 90 degrees to dig the next side.

As the operator nears the fourth corner, the backhoe loader should be offset with the loader positioned outside the chalk line. If the backhoe is kept straight on the chalk line and digging is continued toward the corner, the placement of the loader on or near the bisecting trench could cause loose material to fall into the clean trench, possibly caving in an entire wall.

Once the chalk line has been dug with the backhoe in the offset position as far as the bisecting trench will allow, the operator should fold up the outriggers and set up on the other side of the bisecting trench to a location far enough away so that the outriggers pose no threat to the trench in front of the backhoe. The operator can then continue digging the trench, finishing within a foot of the corner.

Footings

An operator should pull all corners toward the backhoe whenever possible.

4.6.6 Push-back/Pullout

At this point, the operator should situate the teeth a couple of inches from the edge and push the material into the dug area. As enough loose material fills the bucket, the operator should pull the material out and continue this push-back/pullout method until the final line is ready to be taken.

With the bucket in a full open position, the operator should set the bucket teeth directly on the line and sink them into the material. If the trench has been dug right up to the line, then the operator can probably shave the front line in one pass, sinking the teeth to the bottom of the grade and pushing the material back to retrieve it with the bucket rolled over to the flat position.

If there is still quite a bit of material left between the existing chalk line and the trench, the material will have to be moved back bit by bit until the operator can safely shave the last line straight down and finish the footing.

On the last corner, the operator may find a section of material has built up at the edge of one of the bisecting footing corners along the side of the bucket. Taking small bites with the teeth when pushing the material back to clean the last corner will help reduce the amount of buildup. There will always be some material that is unretrievable with the bucket.

If there is a great deal of material left at the corner, the operator can get some of the heavy material into the working area by using the

Chapter Four: Dirt Work

side of the bucket to swing the material over in front of the operator. A shovel can be used to remove remaining dirt.

4.7 Piers

Digging a pier requires removing dirt from a square area that is often no larger than a bucket. A finished pier should be completely square in shape, with straight sides, a flat bottom, and clean sharp top edges.

> *The backhoe must be on a level surface to get a straight side and a flat bottom.*

Digging a clean, square pier means every move the bucket makes must be deliberate and precise, with no mistakes. The key to digging a clean pier is knowing where the teeth are in the ground even when the operator can't see them. Once the teeth have dug past a perimeter line, an irreparable mistake has been made.

Just as with footings, piers need to be chalked or painted out on the material. Operators should extend the corner lines of the pier with chalk or scribe out the lines in the dirt. When scribed, a square pier looks like a tic-tac-toe game. The extending of the corner lines allows the operator to take the line when digging the pier and still have reference points in case extra shaving or squaring of the sides becomes necessary.

Dirt under pressure always seeks the area of least resistance, and the operator needs to make sure the area of least resistance is always in the center of the pier. This can be accomplished by having an area in the middle of the pier that is free of dirt, giving the material a controlled area in which to fall.

Regardless of the type, shape, or depth of the pier, the technique and steps used to dig the pier will always remain the same. The only variables involved are the bucket type and size relative to the pier size.

> *The operator should make sure that all perimeter lines are extended.*

There are buckets made with very specific profiles for digging piers. Although they make the job easier, they aren't mandatory for digging a clean pier; any bucket can be used. What is important is that an operator becomes accustomed to the profile and the feel of the bucket. Using a bucket the same size as the pier is not a good idea. Although the dig would go much faster, it would look like it was done quicker. When a bucket digs into dirt, pressure on the dirt forces it to find the area that has the least resistance. Some of the dirt will go into the bucket because of the force exerted by the bucket cylinder, and the rest will buckle up to the surface. When a bucket the same size as the pier is used, the buckling of the surface will show as broken edges when the pier has been finished. A bucket should be used that has at least 6 inches of clearance from the outside edge of each tooth to the inside pier lines.

4.7.1 Remove Center Material First

After the pier extension lines have been marked on the ground, as much of the center material should be removed as possible without disturbing any of the four lines. If possible, an operator should dig all the way to grade depth. The operator should then shift the backhoe over to either the left or right sideline and set up with the outside tooth just inside the chalk line to make sure the tooth will cut directly parallel to the line. After doing so, the operator should reach back to the rear line and, with the teeth perfectly up and down and right next to the line, cut down and draw forward, stopping about 6 inches short of the front line. As the material breaks loose, it will fall into the hole that was created in the center of the pier. The operator should complete both sides of the pier by removing as much material as possible, while bringing the bottom of the pier as close to grade as possible.

> *If the operator allows the crowd collar to come in contact with the top edge of a pier, the pressure will collapse the side.*

The operator should then shift the backhoe back to the center of the pier, and with the bucket edges in a straight up-and-down position, place the teeth on the inside line of the pier and apply boom pressure directly down. The teeth should be sunk into the dirt 4 or 5 inches and crowded out, kicking the material into the center of the pier.

The teeth should not be pushed deeper than 6 inches on this first inside line pass. The angle that the crowd and boom are positioned in to dig the front line of the pier means that the more the operator pushes the boom straight down without compensating with the crowd arm, the more the front pier line will be damaged.

Every time the teeth are sunk in on the front line and the material kicked back, the operator will need to reach back to the rear pier line and, with the teeth perfectly vertical, remove the material to the same depth as the front. This pull-in/push-back method will need to be continued until the pier is at the desired depth.

To finish the pier, the same pull-in/push-back method will need to be used on each of the side lines. With the backhoe back to one side and with the bucket in the same position as the first time the line was dug, the operator should repeat the process used for the front and rear walls of the pier.

4.7.2 Checking the Work

At this point, the operator should evaluate the pier. The edges must be vertical. If, because of the digging angle or bucket profile, the back edge curves in at the bottom, the operator may need to set up 90 degrees to that side and straighten the edge.

Before the operator moves the backhoe loader to another side to straighten the bottom, the backhoe should be pushed to where the bucket is vertically parallel with the side that curves in at the bottom. If the tips of the teeth can touch the bottom of the pier without the crowd collar touching the front edge, the pier can be finished from that position. The operator should shave the back wall until it is vertical and continue applying pressure until the back wall of the pier is at grade depth.

4.7.3 Retrieving Loose Material

Loose material at the bottom of the pier is difficult to retrieve without the help of someone on the ground. If an operator has help, hand signals should be followed. If an operator is digging alone, extra time will be needed to make sure the pier follows the required guidelines: flat bottom, clean edges, and square sides.

In order to retrieve the loose material, the material must be pulled toward the forward wall. Starting with the bucket at the back of the pier where the rear wall was just shaved, the operator should pull a straight grade toward the backhoe until the teeth make contact with the front wall. (Operators should make sure that it's not the bucket collar that makes contact with the top edge.) After pulling up slightly and releasing the loose material, the operator should bring the bucket back to retrieve the loose material out of each of the corners.

> *When digging more than one pier in a series, an operator should place the dirt where later retrieval will not interfere with the digging of the next pier.*

Once all the loose material is located toward the front wall of the pier, the operator should start at one side of the pier and roll the bucket back to where the bucket floor is resting flat on the floor of the pier. As the operator pulls in toward the backhoe, the bucket can be slowly opened to compensate for the natural reaction of the bucket to curl as it nears the backhoe. When the teeth make contact with the front pier wall, the teeth should be slowly sunk into the front wall and slowly pushed straight back until they clear the wall and are holding the loose material. The bucket can then be pulled out of the pier. This step should be repeated in both corners and, if necessary, in the center.

Chapter Five: Loading

5.1 Loading from a Pile

Whether taking loads from a pile of rock or a pile of dirt, an important word to remember and operate by is *containment*. The moment the loader bucket enters the pile, it starts forcing the material to move. Some of the material will fall into the loader and some will be pushed to the side. This is where containment becomes important.

> *An operator should enter a pile in second gear because it reduces tire spin.*

As material is pushed farther away from the original pile by the loader, more time must be spent retrieving this material. All forward loader passes should be taken pushing the material toward the center of the pile, and as many tasks accomplished in as few moves as possible.

Moving a loader bucket full of material should be a single smooth motion. With the backhoe loader in lower gear, the operator should start at the base of the pile and push the loader toward the center until the machine almost stops. (The operator should not allow it to stop completely because valuable momentum can be lost and the tires may start to spin.) Moving forward slowly, the operator should start by lifting the loader bucket up against the pile and incorporating bucket curl to complete the motion with a full scoop of material.

As the operator is backing out of the pile, there will be an indent where the material was located. The indent is a result of the material being removed, along with the material being forced to the outside edges of the loader.

If an operator took the second load of material from the same spot as the first, the material would continue to spread away from the main pile, increasing the size of the pile base—the exact opposite of the goal.

After the first load is out of the pile, there will be a peak on each side of the indent that was left by the removal of material. Either of

Chapter Five: Loading

After the first bucket of material is taken, all subsequent passes should be taken at one of the peaks created by the preceding pass.

these peaks will be the center of the next bucket load. With the peak centered in the middle of the loader bucket, the operator should start at the tip and enter the pile, curling to take a scoop.

Even though this is the second bucket, the principles of force apply the same as they did with the initial load by leaving two peaks on either side of the loader bucket. Ideally, the operator will work on half of a pile, moving from peak to peak until the pile is small enough to pull the side material in toward the center, creating a smaller, more contained pile.

> *An operator should flatten any spillage as it occurs.*

Depending on the size of the pile and the material it contains, it may be necessary to pull the material back towards the loader in order to keep it in the working area. If the pile is exceptionally large, the operator should use the loader to pull the material from the top down into the work area.

If the pile is small enough, it may be necessary to pull material back almost immediately to keep it in the working area. When this occurs, the operator should go to each side of the pile and bring the ends in toward the center. If the operator continues to use this method while loading, the pile will always stay in a confined area.

5.1.1 Spotting the Truck

Unless an operator is giving instructions to a driver or leveling out a full load on a truck, there is no reason why the backhoe loader should

Loading from a Pile

stop moving when the truck is being loaded. Smooth actions and shortcuts produce speed. When it comes to loading trucks, speed and timing are both needed.

When considering shortcuts in truck loading, proper truck spotting or setup is required to reduce the time the material is in the loader. A slow-moving backhoe loader can traverse all types of terrain without regard to bumps, gullies, or other small obstructions. A fast-moving machine needs a flat, smooth surface on which to operate. When loading more than one truck, the first thing an operator should do is grade the area and continue to keep it flat while loading is in progress. The few minutes it takes to grade an area will be made up in one or two truckloads.

> **SAFETY TIP**
> An operator should never load over the cab of a vehicle.

An important note: Statistically, more backhoe loaders are lain on their sides when working on a flat surface than any other surface because loading trucks is easy and doesn't require constant concentration. Operators, therefore, have a tendency to become neglectful of even the most basic rules.

Operators should never neglect these two rules. First, raising the loader bucket should be carefully timed so it will reach truck height only when the operator is ready to dump. The operator should never travel to the truck with the loader bucket high in the air. The slightest bump could suddenly shift the weight, resulting in the backhoe loader falling. Second, material will be spilled when traveling from the pile to the truck. The operator must make a conscious effort to back-drag and flatten any material that spills, always keeping the working area flat and smooth.

After grading the working area, the operator should position the truck with easy access in mind. The faster the truck can be positioned to be loaded, the faster it can leave so that the next truck can set up. The truck setup area should be as obstacle-free as possible, without infringing on the working area of the backhoe loader.

Because of the need to contain the material, the backhoe loader will need room to work on at least two sides of the pile. Whenever possible, the truck should be set up opposite the pile so that the operator can work from the side. If the truck was set up in front of or in back of the pile, the travel distance would be greater than working from a pile opposite the truck.

Finally, the truck should never be placed in a position where the operator has to raise the loader over the cab to reach the truck bed.

5.1.2 Loading the Truck

While getting the first load, the operator will need to judge the distance that will have to be traveled from the pile to the truck in order to have the loader at the right height when the side of the

Chapter Five: **Loading**

truck is reached. The first couple of loads should be used as trial runs so that the operator can find the point where the loader should start to be raised and, without stopping or traveling with it in the air, have it reach the edge of the truck once the operator is close enough to dump the material. Once this point is found, the rest is a matter of repetition.

The operator should keep in mind that the material will be dumped just before reaching the side of the truck. This puts the distance from the ground to the closest point on the loader lower than it would be if measured from bucket pins on a loader in the full hold position.

If the operator can accomplish all this, then the loader bucket will be opened just as it clears the side of the truck, eliminating the need to raise it quite as high as would have been needed if the operator had waited until the loader, in the full curl position, was opened over the bed.

Having found the raising point, the operator should bring the loader up to the side of the truck and time the dump so the bucket is opened just before the top edge of the bed is reached. The material must not spill until it is over the side of the trench bed.

The timing on this step is important since the operator is trying to save time by advancing the loader opening. If the timing is off, the material may be spilled before the loader reaches the side of the bed.

Once the lip of the loader has cleared the side and the material has started to fall from the bucket, the material should be dumped as quickly as possible. (An exception to this is when rock or other heavy material is being loaded.) The operator does not need to stop and

An operator should be aware of the laws in the area regarding load height and weight.

wait for all the material to fall from the loader. Once the loader is at about the three-quarter dump position while still traveling forward, the operator should continue to dump while placing the backhoe loader in reverse and starting to back away. The rest of the material should fall out by the time the reverser engages and the machine actually starts to move. The instant the loader bucket edge clears the side of the truck, the operator should drop it as quickly as possible to lower the center of gravity and retrieve the next load.

5.1.3 Leveling the Load

When loading a truck with a loader, the material tends to pile in the center of the truck's bed, leaving voids along the sides. One solution is to take the last load of material and use it to flatten the material in the truck.

The operator should put the loader in the full curl position, raise it high enough to clear the side of the truck, and stop. With the loader bucket in the air and at the side of the truck, the loader bucket should be slowly rolled so that the lip of the bucket will fall open just inside the bed of the truck.

> *In some states, the load is the operator's responsibility. An operator should know the laws in the area where the work is being done.*

The operator should then creep forward while opening the loader to empty the contents against the inside bed wall. When the loader bucket is in the full dump position, it should be raised until the cutting edge is at the same height as the side of the bed. The operator should then drive slowly forward until the nose of the backhoe meets the side of the truck bed.

After lifting the loader bucket to bring the cutting edge out of the material, the operator should curl the bucket to the flat position, lower it, and place a little pressure on the load. This will flatten the load.

When time permits, the operator should turn around, set up, and clean up the load with the backhoe. However, this type of leveling should be used for bulky loads or loads that are off-balance in the bed.

> *Dumping a layer of light material in the bed of a truck before loading heavy or bulky material is called bedding a truck.*

Covering the load is becoming more prevalent as more laws are passed to protect the paint on the nose of passenger cars. When loads are being tarped on a project, the operator may need to take a little more time on the leveling process to assure that the tarp will cover the load.

5.1.4 Tips

When loading heavy, bulky material such as large pieces of concrete or rocks, the operator may be unable to set the piece in the truck bed gently; it may have to fall from the bucket. When this occurs, the oper-

Chapter Five: Loading

ator should get a loader full of clean material such as sand or road base (they work well because they will not interfere with a recrusher if the load goes through a recycling plant) and lay it in a pile in the bed of the truck. Now the operator can safely drop the material in the truck bed without the material going through the bottom of the bed.

If there is ever a question as to the load's weight, it is always better to be safe than sorry. An operator is better off loading the truck on the light rather then getting an overload ticket.

Different states have different regulations and laws governing trucks and their payloads. In some states the driver is responsible for the load being carried; other states have laws specifically pertaining to trucks that are used to haul material to and from a construction site. These laws place the responsibility on the operator or whoever is loading the haul truck. It just takes a telephone call to the local highway patrol to find out about the laws regulating load responsibility in the operator's area.

5.2 Loader Grading

5.2.1 Loader Bucket Edges

There are many preconceived notions about loader grading. One is that the only usable edge on the loader bucket is the cutting edge on the front lower floor. Actually, the loader bucket is equipped with many sides and edges that can be incorporated into loader grading, and the operator will find that each of the different sides and edges can serve a very specific grading purpose.

A loader bucket is equipped with many usable sides and edges.

The angle at which the loader bucket meets the ground is in direct correlation to the angle of the rear tires as they rest on the ground. This formula is the basis for all loader grading; it means that the rear tires of the backhoe loader have to be on a level surface during the entire grade pass in order for the loader bucket to cut a level grade. This can be done either by trimming the grade or by shifting the distribution of weight from one side of the backhoe loader to the other by using the boom.

An operator will be able to do grading tasks quickly and efficiently by learning to use the different parts of the loader bucket and by learning to use the constantly shifting weight of the backhoe to an advantage.

The strongest and most versatile of the loader bucket edges is the front lower edge. Depending on the angle at which the edge meets the ground, it can cut heavy material, light material, trim high spots traveling forward, and fill low spots traveling backward. Learning to use the loader well enough to trim specific high spots quickly and with a minimum of passes sets the precedent for all loader grading.

5.2.2 Operator Methods

Once an operator acquires the ability to manipulate the loader bucket in such a way that it actually does what an operator wants, all other loader work becomes a question of one's method of approach.

A loader bucket reacts to the material it is cutting. When the cutting edge is set at a particular depth in the grade and as the backhoe loader starts to travel forward, the natural tendency of the cutting edge is to continue cutting, not at the depth at which the cutting edge was originally set, but at the angle at which it is set in the grade.

Knowing how the loader will react to the material when the cutting edge is lowered to the ground is knowledge gained through experience, and the most efficient way of gaining this experience is by watching. Inexperienced operators should watch what happens when the loader bucket makes contact with the ground. Do the front tires start to squat, indicating that there is too much loader pressure? Do the rear tires sometimes lose traction, again indicating too much loader pressure? Is one side of the loader digging deeper than the other?

These are all circumstances that arise during grading, and they are all detriments to a good grade. Once an operator learns that everything must be compensated for, the experience can all be added to an operator's bank of grading knowledge.

In this chapter, a loader bucket in the level position with the bucket floor parallel to the ground will be referred to as a loader bucket in the 3 o'clock position. A loader bucket in the full dump position, with the loader floor perpendicular to the ground, will be referred to as a loader in the 6 o'clock position. Any position between these two hours will be numbered accordingly as 3:30, 4:30, 5:30, etc.

5.2.3 Grading Guidelines

A few guidelines can help an operator when loader grading. First, before any grading takes place, the operator should set the backhoe loader on a flat surface, such as asphalt or concrete, in the 3:30 position and slowly lower it to the ground. It is important that both sides of the loader bucket meet the ground at exactly the same time. Any discrepancy in bucket distance to the ground can be made up either by adding air to the corresponding rear tire or by taking air out of the tire that corresponds to the loader side.

Second, it is easier to fill than it is to cut. When trimming a high spot from the grade, it is easier and more time-efficient to overcut the spot slightly with the loader and then backdrag material to the area with the rear edge of the bucket.

An operator should not be afraid to undercut the grade slightly, because it is easier to fill than it is to cut.

Third, when working in deeper or harder material, it will be necessary to make a gouge to give the loader bucket a place to start. Instead of turning

Chapter Five: Loading

around and cutting a section out with the backhoe, the operator should swing the boom to the side being cut and set the loader in the 4 o'clock position. The operator then needs to get the low corner of the loader bucket to bite first. Once that corner of the bucket is in the ground, the weight of the material filling it will pull it deeper into the grade. If the material is very hard, the operator may need to gouge out one pass and move over a couple of feet and repeat the operation.

Fourth, when fine-grading with the loader, the operator should pay special attention to the cutting edge during a turn. The backhoe section of the machine is very heavy and its center of gravity is relatively high. These two factors together act against a backhoe in the form of forward momentum when a turn is made under power. When a right turn is made with the cutting edge skimming the ground, the left corner of the cutting edge will start to cut material. The same thing occurs when a left turn is made.

To overcome this problem, the operator can compensate by swinging the boom to the same side in which the loader is turning, or the operator can simply not turn while fine-grading. All cut passes should be made in a straight line.

5.2.4 Grading with a Raised Perimeter

Before any large-scale grading can take place, the backhoe loader must have a level area from which to start. At the minimum, this area needs to be slightly longer then the length of the backhoe, from the back of the bucket to the tip of the cutting edge, to allow for some extra room.

> *The closer the operator can get the edge of the loader bucket to the raised perimeter, the less material will flow over or on top of it.*

If an area is large enough to warrant a backhoe loader being brought in to grade it, it probably has a defined or raised perimeter. (A raised perimeter is anything from which the operator can visually transfer a point to the loader in order to maintain a constant grade.) It could be anything from a form board, if grading a large slab, to the existing asphalt, if grading down a street section or dig out.

If the area being graded has a defined perimeter, then the operator should start at the side with the deepest cut. If the cut in the grade is excessive, it is worth the operator's time to set up and grade the area flat with the backhoe bucket. If the cut in the grade is minimal, the backhoe should be lined up to the side chosen as the starting point.

5.2.5 Six o'Clock

With the backhoe parallel to the starting side and as close as possible to the raised perimeter without touching, the bucket should be rolled to the 6 o'clock position. With the backhoe loader in low gear, slowly moving forward, the operator should gradually lower the loader

Loader Grading

Grading to a raised perimeter.

bucket until it makes contact with the ground.

Manufacturers do not recommend grading with the loader in the 6 o'clock position because of the possibility of damage to the front loader ram. When grading with the loader in a rolled-over position, the operator should not allow the material to force the loader past the 6 o'clock position. If too much material accumulates, or if the material is too hard, the loader should be changed to the 5 o'clock position to cut the heavy material, and then returned and fine-graded with the loader in the 6 o'clock position.

> *An operator should never start a pass with the loader bucket already in the grade.*

An operator should never start a loader pass with the loader already on the ground, because this will result in a gouge that the rear tires will have to cross. As they do, the tires will dip and this dip will be transmitted to the loader bucket in the form of a hump. The operator will then have to go back and fill it later.

Once the loader bucket edge makes contact with the ground, the operator should watch what is happening to the material being shaved. If the correct amount of material is being shaved in a single pass, the high points in the grade will be pushed into the low spots and the excess material will spoil out the side of the loader and create a windrow. But unless the grade is extremely close, or sand or loose base material is being graded, chances are this scenario will not occur.

Chapter Five: Loading

5.2.6 Loader Floating

A few feet after the loader bucket has made contact with the grade, the material being shaved will build up too rapidly for it to be pushed out the side of the loader into the windrow. This buildup will cause the loader to start floating on the material already cut until all of the shaved material has passed under the loader, at which point the loader will slowly return to the grade, leaving a hump.

> *An operator should never hesitate to use the steering brake to assist in a turn.*

There are two solutions to this floating effect.

One is to simply change the angle of the loader, being careful to make the change in very small increments. The operator will need to find a point between the moment when the loader wants to gouge the grade and the moment when the loader, filled with material, still wants to float.

The second solution is to pick a point on the loader bucket and align this point with one on the raised perimeter. During the grading, these two alignment points should be used to keep the loader at a constant grade. When the loader shows the hint of raising, the wheels should be turned to push the shaved material toward the ungraded section. The operator should avoid raising the bucket until the inside edge of the loader has cleared the outside edge of the area being graded.

By keeping the bucket on the ground until the inside edge has cleared the edge of the pass, as the loader is turning, the outside edge is continuing to grade the section next to the one being worked on. When it is time to line up in order to grade the adjoining section, the operator will have some good grade pads to grade by.

When it is time to re-enter the grade to continue with the pass, the

When a section of grade requires a change to the 5 o'clock position, the technique used to grade will also have to change.

operator should remember to start back far enough for the entire backhoe, from the back of the bucket to the tip of the cutting edge, to be in the already graded area. The operator should ease the backhoe forward, allowing the loader to drop until it has met its alignment point or the existing grade. This process should be repeated until grade has been achieved along the perimeter side.

5.2.7 Hopping

If, with the loader bucket still in the 6 o'clock position, an operator encounters very hard material, or if a great deal of material must be cut in a single pass, then the loader may start to bind because it is unable to move the material out of the way or cut through it. When sufficient pressure has built up on the loader, it will quickly release and cause the loader to start hopping. This hopping can be stopped by either taking smaller bites when lowering the loader bucket into the grade, or by slightly changing the bucket angle.

5.2.8 Five o'Clock

In the 6 o'clock position, the loader bucket places the cutting edge in a harmless position to the grade. The loader does a good job of grading lighter material, but when it comes to cutting hard material or larger high spots, the angle of the loader needs to be changed.

When a section of grade requires the operator to change the loader to the 5 o'clock position, the grading technique will also need to change. When the operator places the loader bucket into the grade, it is important that enough weight remain on the front tires to make steering possible. It allows the operator to adjust the loader for the drag created by the extra weight in the loader.

With one side of the bucket matching a flat section of the grade or matching an alignment point on the raised perimeter, the operator should travel slowly forward, allowing the outside edge of the bucket to cut away the high material. At the same time, the inside of the loader bucket should either match the existing grade or align to the point on the raised perimeter.

> *The harder the steering brake is applied in a turn, the more the outside cutting edge will dig into the material when the loader bucket is in a cut position.*

It is easier to remove high spots if the operator has a point relatively close to grade so that one side of the loader can be run and the other side of the loader can be used to cut the high spot. Whenever possible, the operator should balance the opposite edge of the loader bucket between the point close to grade on one side and the high part of the grade on the other side.

As the cut material fills the bucket, the added weight will draw the cutting edge deeper into the grade, causing either gouging or the re-

Chapter Five: Loading

moval of too much material. To offset the tendency of the loader to lower when being filled, the operator should be prepared to compensate immediately for the loader drop by slightly lifting the bucket as the cutting edge starts to cut the high spot. The loader bucket should continue to be raised slowly as the cutting edge progresses into the material. The amount needed to raise the loader to compensate for the drop depends on the drop as gauged by the alignment marks on the raised perimeter, or by monitoring the distance between the cutting edge and the ground on the side of the loader bucket that is skimming just above the area that is close to grade.

It is not necessary to continue cutting when the material starts to spoil out of the side of the bucket. The more material being pushed by the loader, the more difficult it will be to make and maintain grade. The operator should exit the grade area often, keeping the loose-cut material in the ungraded area.

5.2.9 Three o'Clock

Grading with the loader bucket in the 3 o'clock or flat-loader position is usually reserved for skimming small bumps off an already flat area or for grading lighter material such as sand or base.

In addition to a change in loader bucket position, there will be a change in the way the loader reacts to the material being graded; the loader bucket in the 3 o'clock position will hold more material then when placed in the 5 o'clock position. The added material means ad-

The wear pattern of the cutting edge may cause an operator to open or curl the loader bucket slightly to achieve a correct 3 o'clock setting.

ditional down force on the loader because of the extra weight.

In flat-loader grading, the extra down force will not drive the cutting edge of the loader into the grade as it would if the loader was in the 5 o'clock position. However, it does create other obstacles. One involves working with the material itself. When an empty bucket in the 3 o'clock position is run across a grade of lighter material, the loader will float across the surface regardless of how much down pressure is applied. The solution is to fill the loader bucket with material and slowly open it to the 3 o'clock position. As the loader is run across the same light material, the material placed in the loader on the next pass will add enough weight to cause down force, driving the cutting edge into the grade.

> *The wear pattern of the cutting edge may cause the operator to open or cut the loader bucket slightly to compensate for a correct 3 o'clock setting.*

A second obstacle is in emptying and filling the loader bucket. When flat-loader grading light material, it may seem that most of the operator's time is spent compensating for the change in loader weight. The solution is an easy one and can be applied to flat-loader grading of all types of material. When making the first pass, or when the first loader bite is taken from a high spot, the operator should keep about half a bucket full of material. This material should be held in the loader as flat-loader grading is continued.

Once this technique is acquired, the operator will discover some distinct advantages. When skimming bumps off a grade, for example, loose material will fall from the loader and fill any low spots. When grading, the added material in the loader will stabilize the entire backhoe loader, making the grading easier and quicker.

5.2.10 Float Position

The backhoe loader may be equipped with a float position built into the loader valve assembly.

This position is actuated by pushing the loader control lever all the way forward until it locks into position, permitting the hydraulic oil to flow uninterrupted through the valve into the front loader cylinders and back again into the valve. This uninterrupted flow of oil allows the loader to ride along the surface of the grade. The weight of the loader alone should supply enough down pressure to the loader's lower rear edge to trim small high spots and fine-grade lighter materials.

A backhoe loader set with the loader in the float position can only work material while traveling in reverse. If an operator tries to work the material while traveling forward with the loader set at float or any position past 3 o'clock, the cutting edge of the loader will dig into the grade.

A loader bucket set to float relies on the rear lower edge on the back of the loader bucket to accomplish the work. As with any loader grading, a change in loader angle determines the amount of material being

Chapter Five: Loading

The float position is used when traveling in reverse.

cut. Curling the loader even slightly will dramatically change the effect of the rear cutting edge on the material. These slight changes in the loader bucket position can be used for specific grading procedures.

When the loader is set flat with the loader floor parallel to the grade, the operator is in a position to smooth the grade. This position works very well for putting the finishing touches on a graded area or for spreading lighter materials such as sand or loose dirt. Curling the bucket even slightly from the flat position forces the rear of the loader bucket to act as a cutting edge.

When the operator is ready to put the finishing touches on the grade, all of the backhoe loader weight should be placed on the front tires, the loader control pushed to the float position and, in low gear, the reverser lever engaged to start traveling backward.

When polishing a graded area, the operator needs to watch three areas: the angle of the rear bucket edge where it meets the ground, the material after it has passed under the bucket, and behind the backhoe because the float position is in reverse.

The type of material being polished has everything to do with the angle of the loader bucket when it makes contact with the ground. In most cases, it is a matter of trial and error to get the consistency desired.

5.2.11 Grading Import Material

Grading loose import material is like grading native material. The two major differences are that the operator has complete control over the amount of material being used to fill the grade when using import material, and properly compacted import material will have an even density when the bucket edge is cutting through it. Native material, by contrast, can have variations of hard and soft spots that can cause the loader to rise and dip, making the job more difficult.

When grading import material, operators need to remember that before any loader grading is to take place, the material must be compacted, including adding moisture if necessary. Also, the import material should be overfilled and compacted and then cut to grade. Operators should not try to lay the material in at grade right from the start, because after it is compacted, the operator will have to go back and fill in the low spots. Finally, the operator should have a grading plan in mind before bringing any material into the grade. Whether the material is arriving by loader or truck, placement is the operator's responsibility. It is easier and more time-efficient to start at one end of the grading area and work to the other side instead of starting at the center and working out.

> *When fine grading very large areas, some operators think removing the backhoe bucket gives them steadier and more responsive loader bucket control.*

5.3 Backfilling

5.3.1 Introduction

It can take a great deal of time to backfill a trench properly. A tight, compacted grade can eliminate many of the problems associated with a loose trench. If an operator takes the time to backfill a trench correctly, then there is no worry about failed compaction tests, and about rear tires sinking into a loose trench while loader grading, throwing off the loader bucket grade.

> *An operator should mark all ground obstacles before backfilling begins.*

5.3.2 Compaction

If high percentages of compaction are a concern, the operator will need to pay close attention to the height of the lift and the amount of water added to the backfill material. The operator will need to limit the amount of material put in the trench at one time. Putting in less material than is required to fill the trench is called a lift.

The operator will also need to monitor the amount of water added to the backfill material to make it more compactable. The optimum water content is the amount of water added to a specific type of material which helps it to compact to its greatest density. The amount of water varies depending on the type of material. Too much water makes the trench a muddy mess with no compaction value whatsoever, and too little water means the operator will have to go back, process the material with more

> *If the cutting edge doesn't line up with the asphalt or concrete when backfilling on a street, the operator should swing the boom to one side or the other for compensation.*

Chapter Five: Loading

water, and recompact the trench.

Being familiar with the type of material being used for backfill and knowing how much water it takes to bring that type of material to optimum water content is beneficial and can save a great deal of time. Although much of this knowledge is acquired by experience, a rule-of-thumb is that adobe soils require less water and more processing, and sandy soils require more water and less processing.

5.3.3 Adding Water

The procedure used to add water to the backfill material depends on the accessibility of the water, the length of the trench, and the percentage of compaction needed.

The quickest way to add water to the material is to have a water truck wet the material while it is still in spoils on the side of the trench and then process it as it is laid in. Another method is to have someone with a water hose spray the spoils as the material is laid in the trench with the loader. Both of these methods work when the material needs to be tight but not high in compaction percentages.

> *It is easier for an operator to add water to dry material then it is to dry wet material.*

Where high-compaction percentages are a factor, the operator will need to work harder to achieve the optimum water content. The key to getting optimum water content is to make sure the water is distributed evenly throughout the backfill material. The best way to accomplish this is to spread the spoils out with the loader and add the water while thoroughly mixing the material. One way to do this is to have whoever is watering the material set up in a place where there will be no interference with the loader work while still being able to reach the working area with the water. The more room the operator has to work, the quicker and easier the work can be accomplished.

> **SAFETY TIP**
> On softer material, an operator should make sure the side of the trench can support the weight of the backhoe loader before making parallel passes on the top edge.

If the trench the operator is preparing to backfill is exceptionally long, it should be divided into smaller, more workable sections, and the operator should work the water into the material a section at a time. Ideally, the water should be sprayed as evenly as possible over the spoil area. If too much water accumulates in one spot, the operator should lay dry material from another section on top of the wet material and mix it in.

5.3.4 Using Tire Imprints to Gauge Moisture

A fairly accurate way of determining the correct amount of water that needs to be added to the material is to watch the tire imprints as the material is mixed with the loader. As the material is pushed forward, material that is too wet will show the imprint of the entire tire and leave a raised track where the wet material squished up the sidewall of the tire. The area may puddle even after the backhoe runs over it, leaving a low spot. Material that is too dry will leave either a clean track with crumbled edges or a clean track that crumbles under the slightest of pressure. The ideal tread mark will have a crisp imprint of the tire with little or no sidewall marks, and the tread mark will hold together under pressure and not crumble.

5.3.5 Lifting

When the operator has the desired water content, it is time to prepare for the first lift. If wheelrolling is the means of compaction, the operator should try to lay the material more evenly and avoid highs and lows. Using conventional wheelrolling, a backhoe loader cannot properly backfill a trench more than about 16 inches deep if good compaction is desired. If an operator tries to wheelroll a trench that is too deep, either the backhoe loader will be placed in a position where the operator will need to push it out, running the risk of forcing dirt between the rim and the bead of the tire and possibly causing a leak, or the steering cylinder ram will be damaged by being scraped along the top edge of the trench.

> *The thinner the lift, the tighter the compaction.*

The deeper the trench, the more weight that is transferred to the low side of the backhoe because of the shift in the backhoe's center of

Putting less material in a trench than is needed is called a light.

Chapter Five: Loading

gravity. The goal is to place as much weight as possible on the trench to achieve good compaction while keeping the rims and steering cylinder in one piece.

If the trench is over 16 inches deep and there is no stomper or vibratory plate on the job, the operator will need to lay in the first lifts of material straddling the trench with the backhoe and then tamp in the lift with the back of the bucket. If the trench contains plastic conduits, the operator should be careful not to tamp so hard as to crush or disfigure them.

> **SAFETY TIP**
> When backfilling on public roads, an operator should be sure to establish a routine of signals with the laborer.

When the first lift has been tamped, the operator should push the backhoe loader off the trench, lay in another lift, and reposition the backhoe over the trench to tamp in the second lift. This procedure should be repeated until the trench has been backfilled high enough to allow wheelrolling.

5.3.6 Repetition

When backfilling, many moves will be repeated so operators will need to determine the pattern and rhythm that is easiest and quickest.

The quickest way to lay a lift of material into the trench is to place the loader bucket in a 5 o'clock position and run parallel to the trench, allowing the material to flow into the trench in an even lift. Generally, a lift of 6 to 8 inches is considered to be a good compactable lift. The operator will be able to adjust the amount of material going into the trench by raising or lowering the loader bucket.

If material entering the loader bucket is pushed out of one side into the trench, then an equal amount will be flowing out of the other side, creating a windrow of material along the outside of the loader bucket. The only way to get this windrow of material back in working range is to set up the backhoe to retrieve the material at a 45-degree angle the entire length of the trench.

Running the backhoe parallel to the trench in order to lay in the first lifts and for final cleanup is acceptable practice. But from the point of the first lift until the finished filling, the operator should avoid the outside edge of the spoils and the time it takes to bring it within working range.

5.3.7 Wheelrolling

When the trench is ready to be wheelrolled, the backhoe loader should be straddled over the spoils with the loader bucket in the 6 o'clock position. The backhoe loader should be run over the spoils while the loader is lowered just enough to knock the tops off, allowing the material to flow down the sides of the spoils and create a flat windrow at the bottom. This will provide a flat surface for the two outside tires to ride on as the operator works the material on the out-

If possible, an operator should have material in the loader bucket while wheelrolling.

side edges of the trench.

After the first lift of material has been laid in, the trench should be wheelrolled with one quick pass. The loader bucket should then be filled with material from the spoils and the area wheelrolled again with the added weight. The loader bucket should be as low to the ground as possible to eliminate the possibility of laying the backhoe on its side.

This added weight in the loader bucket will need to be repeated after every lift. After the first lift has been thoroughly wheelrolled, the operator should retrieve the outside windrow that was created when the first lift of material was put in. Working with the backhoe at a 45-degree angle to the trench, the operator should push the material next to the trench within working range for the next lift.

5.3.8 Gauging Material Flow

After the outside windrow has been brought closer to the trench, the loader should be lined parallel to the trench and the loader bucket placed in the 4 o'clock position. Placing the loader bucket in the 3 o'clock position would create a low spot in the trench that would have to be filled later. The material should be pushed forward into the trench. The operator should watch the outside of the loader, observing when the material starts to flow outside of the loader away from the trench. As soon

> *If, while backfilling, a loader bucket catches a manhole, the backhoe loader will stop.*

as this starts to happen, the operator should turn the steering wheel and apply the turning brake in the direction of the trench. If too little material is flowing into the trench, the turning brake should be applied harder, causing the loader to approach the trench at a greater an-

Chapter Five: Loading

*A change in the loader bucket angle or
the steering brakes can increase or decrease material flow.*

gle. If too much material is flowing into the trench, giving too thick a lift, then either the loader should be raised enough to release some of the material from the loader, or the opposite turning brake should be applied and the backhoe turned parallel to the trench until the material starts to spoil out the opposite side of the loader.

After a pass is completed, the operator should give the trench another good wheelroll with material in the loader bucket by backing up parallel to the trench and starting at the newly created spoils.

This process should be repeated with the operator cutting in harder to the trench when the material starts to spoil out the opposite edge of the loader bucket.

5.3.9 Side-cutting

Straddling the trench to pack the material in with the back of the bucket is a very efficient way of compacting the first few lifts of a deeper trench. However, when time is short, side-cutting is a good alternative. Instead of bringing the material in the trench up to top to wheel-roll it, side-cutting takes one side of the trench down to the depth of the first lift of material.

Operators should start at one end of the trench and run the loader bucket in the 6 o'clock position over the tops of the spoils the entire length of the trench, knocking off just the tops. Taking a full loader bucket of material to the starting point on the trench, the loader should be lined parallel, with the loader edge resting on the edge of

Backfilling

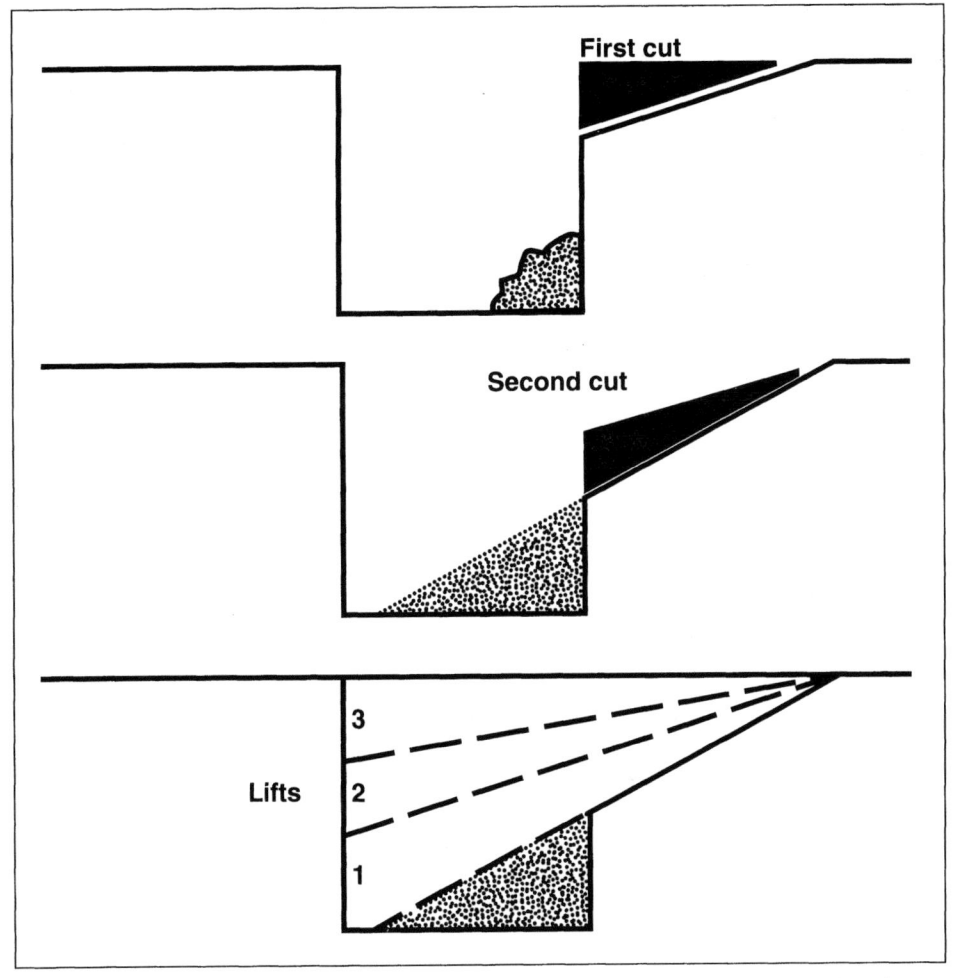

An operator should take note of angle cuts and the backfilling angle when using the side-cutting method.

the trench. The operator should then change the loader bucket to the 3 o'clock position and push the full loader into the trench spoils and allow the material to spill the entire distance into the trench.

Once the operator has reached the far end of the trench, the loader bucket should be emptied and turned around. With the front tires directly at the center of the trench, the operator should slowly drive forward, allowing the tires to wheelroll the trench. An operator needs to pay special attention to the tires at this point because they are doing the actual compaction.

As the backhoe loader continues along the trench, the height of the first lift should slowly start to decrease until eventually the trench is at its original depth. The trench should be wheelrolled until the back-

Chapter Five: Loading

hoe will not go any farther in the trench. At this point, the operator should back out the same way that was entered.

The operator should then set up at the beginning of the trench with the loader set to the far edge of the trench and in the 3:30 position. After slowly re-entering the trench, the loader bucket will cut a nominal amount of material until the rear tire enters the trench upon which the loader bucket angle will change dramatically and immediately. The operator should use the change in loader angle to cut the ridge off the top edge of the trench, allowing the spoiled material to flow either into the trench or out the sides.

> *Because of the backhoe's angle when backfilling a section that has been side-cut, there is tremendous pressure on the beads of the tires. The operator should be sure the tires are fully inflated or slightly overinflated.*

The operator will need to cut as much material as possible while keeping the backhoe loader moving. The operator should continue cutting the edge down using the same angle until it is not possible to wheelroll any longer.

The operator will be able to tell when the loose section of the trench is coming up because it will either look loose or it will be substantially lower in depth. When this point approaches, the operator should change the angle of the loader to the 4:30 position and continue pushing. As the loader bucket position is changed, the operator should try to not change the cut on the top edge of the trench. Changing the loader bucket position will push more material into the trench in front of the loader, allowing wheelrolling to continue without stopping.

With a rough cut completed, the operator should turn around and repeat the cut one more time, with the loader in the 3:30 position, from the opposite direction. This second pass cuts deeper on the edge and a little deeper in the trench. When the far end is reached, the operator should take a full load of material and, with the loader bucket very low to the ground, wheelroll the trench very slowly.

Summarizing, the first cut of 12 inches brought the trench down to 24 inches. The second cut brought it down to about 20, with another 3 or 4 inches of the trench being cut by the opposite end of the loader, bringing it to a 16-inch compactable section.

With a loader full of material, the trench should be wheelrolled a few more times, laying in material to bring the V-cut back up to grade.

Wheelrolling can be very hard on tires and rims. The tires must be fully inflated and, when turning in and out of the trench, the steering brakes should be used to assist the front tires.

5.3.10 Backfilling at 90 Degrees

There are times when the trench just needs to be filled. If the trench is a test hole in the middle of a wheat field, just filling the hole should be fine. But if the trench is crossing an often-used dirt road, an operator should think twice about just covering it up. An operator should know that if the trench is just covered up, there will be a divot in the road in just a couple of hours.

> *Poor workmanship will always come back to haunt an operator.*

Instead of just pushing the dirt into the trench, the operator should approach the trench at a 90-degree angle with the loader bucket in the 3 o'clock position, pushing the material into the trench. There is no need to move the loader from the 3 o'clock position until it is time to wheelroll. If there are any low spots, they can be filled after the rest of the trench is full.

Starting at one end of the trench and working one loader bucket width at a time, the operator should fill the entire trench. Once the trench is full, all low spots should be filled in. By now the trench should look like a mountain range in the middle of the road.

With the loader bucket empty, the trench should be wheelrolled and the loader then put in the 3 o'clock position. The operator should position the machine at the beginning of the trench, with the loader splitting the trench in half. With the loader bucket on the grade, the backhoe loader should be taken down the middle of the trench, allowing the material from half of the trench to fill the loader bucket. (When the backhoe loader gets both tires into the trench, it may sink a little. The operator can solve this by slightly raising the loader bucket.)

When the far side of the trench is reached, the operator should curl the loader bucket to a holding position and turn the machine around. After making the last pass, the operator has accomplished four things: another pass was made with the tires in the trench, helping compaction; a flat area was made on the trench for the backhoe loader to travel on;

Backfilling at a 90-degree angle.

Chapter Five: Loading

with the loader bucket in the 3 o'clock position, some low spots were filled; and the loader bucket was filled with material for the next pass.

With the loader bucket lowered close to the ground to lower the center of gravity, the trench should be wheelrolled again, right down the center. The extra weight from the full loader bucket should compact the grade even further.

After the pass is complete, the operator should turn around and, with the loader bucket back in the 3 o'clock position and on the grade, the loader should be run down the opposite side of the trench, knocking any remaining high spots into the low spots.

To fill any remaining low spots, the operator should dump the material from the loader at the end of the trench. After raising the loader high enough to clear the pile just made, the operator should travel forward sufficiently so that when the loader bucket is set back to the ground, it is done at the far edge of the pile. With the backhoe loader turned around, the operator can backdrag the material down the center of the trench, bridging the material. Bridging the material creates a compactious crust on top of otherwise soft material, giving the illusion of full compaction.

5.3.11 Backfilling at 45 Degrees

Backfilling a trench at a 90-degree angle is an ideal way to get material into a trench in a hurry, but it never allows the operator the convenience of watching the material enter the trench. If the operator is gauging material flow for a lift, that convenience is necessity. For example, it is necessary when an operator is receiving hand signals from a laborer who is standing in front of the backhoe in the trench, holding a riser.

Backfilling at a 45-degree angle will give the operator that convenience. For example, a trench is 2 feet deep and it has to be brought up in lifts. There is a block wall running the length of the trench, 3 feet away on the opposite side; all of the spoil material is on the opposite side of the block wall.

If room permits, the first steps will be the same as with any other type of backfilling. The backhoe loader should be parallel to the trench, with the loader bucket in the 6 o'clock position over the top of the spoils. A loader full of material should be taken from the back side of the spoils and lined parallel next to the trench. With the loader in the 3 o'clock position, the operator should run down the edge of the trench and allow the full loader to push a lift of material into the trench. If the lift becomes too thick, the operator should raise the loader until the lift thins out.

After the operator has wheelrolled, added water, and finished the lift, the backhoe loader should be lined up at a 45-degree angle at the beginning of the trench, far enough back to take all the material in that

Backfilling at a 45-degree angle.

pass. The operator should then set the loader bucket to the 4 o'clock position and push the material toward the trench.

If an operator is just pushing the material into the trench without regard to compaction or lifts, the operator should continue pushing the material into the trench until the front tire makes contact with the edge of the trench. At that point, the loader should be raised and the backhoe loader lined parallel next to the pass just completed.

Looking at the area where the pass was made, the operator will see that the outside edge of the loader wasn't able to push the side of material into the trench, leaving a hump on the edge of the trench. This was caused by the 45-degree angle not allowing both tires to approach the edge of the trench at the same time. When lining up at a 45-degree angle for the next series of passes, the hump must be incorporated into the trench.

If an operator is not simply pushing material into the trench, and the amount of material flowing into the trench has to be taken into account, the backhoe should be lined up at a 45-degree angle to the trench, with the operator making certain the outside of the loader will take the residual of the spoil material as the backhoe loader progresses toward the trench.

With the loader bucket in the 4 o'clock position, the operator should start pushing the material toward the trench, watching as the loader bucket makes contact with the spoils and continuing to watch as the material starts to enter the trench. As the operator is pushing, the loader bucket must be just touching the ground, with the front

Chapter Five: Loading

tires absorbing the weight of the material.

The instant the material starts to fall over the edge of the trench is the instant the operator needs to react. The operator should start to turn away from the trench, not waiting until the material has reached the proper lift height before starting the turn. With the loader bucket in its present position, turning and lifting the loader will increase the amount of material flowing into the trench. Consequently, turning must be started just as the material starts to fall.

If too much material is flowing into the trench, the operator should lift the loader bucket *while still turning* to slow the flow; the operator should use the steering as much as possible to control the flow. The operator should continue running the backhoe parallel to the trench until the flow of material starts to decrease. The loader should be raised and the backhoe loader lined up at the starting point next to the pass just completed. The loader bucket should be lowered to the ground and the second pass started by using the small windrow caused by the first loader pass to align the backhoe.

Just as in the first pass, the operator should time the start of the turn to coincide with the beginning flow of material. The operator should turn away from the trench, running parallel until there is no more material. The process should then be repeated.

There are several benefits of using this technique to backfill. First, all backhoe motions are toward the trench or toward one end; there are no wasted movements. Second, the material is kept within a confined area and so there is no need to take time to retrieve outside material and bring it back into the working area. Third, since the operator is starting at one end and working toward the other, the loader bucket is always riding on a flat surface, saving the operator from having to go back and grade.

5.3.12 Finishing

When the backfilling is nearly completed, the operator should start accumulating the rest of the backfill material and spreading it evenly along the top of the trench. This extra material should be wheelrolled until it is well-compacted.

Starting at the beginning of the trench and with the loader bucket in the 3 o'clock position, the operator should center the backhoe loader on the trench in a straddling position and, with the loader, cut it flat, allowing the excess material to fill the loader bucket. When the excess material starts to come out of either side of the loader bucket, the operator should curl the loader bucket and dump it in a stockpile area. The operator should return the empty loader to the point where it was last curled, lay the bucket flat, and repeat the process until all of the loose material is either in the trench or in the stockpile.

Chapter Six: Utilities

6.1 Locating Utilities

At some point an operator will be expected to locate an underground utility. Whether by digging on an actual utility mark or by digging slowly with the possibility of a utility being in the area, an operator will be expected to find a utility without causing any damage.

Having a general knowledge of utilities will help in identifying the type of material the conduit is made of and what utility is inside. Once this information is known, the operator can make decisions concerning locating the utility in the safest manner possible.

It is important that the operator be aware of the laws and regulations in the area before work begins. There is a one-call service available throughout the United States. Operators are urged to use the service before digging. Certain states have strict laws regarding underground utilities, and it is a federal offense to dig over an existing fuel transport line without notifying the utility company. Other laws state that under no circumstances will a utility be located with a motorized piece of equipment.

> *Before digging begins, an operator should know the laws in the area regarding underground utilities.*

6.1.1 Feel

Locating utilities successfully requires the operator develop a feel for the machine. The word *feel* in this context is used to describe the operator's ability to sense any changes in density or pressure in the material being excavated. This ability is something that must be acquired through experience, patience, and practice.

Feeling for an underground utility is a constant

> *An operator should keep the bucket moving at the same speed and take the same amount of material on each pass.*

Chapter Six: Utilities

comparison, from one instant to the next, of how the material pulls when the bucket is being drawn through it. Consequently, it is important for the operator to take the same amount of material on each bucket pass and to keep the bucket moving at the same rate throughout the pass until the utility is located.

Practice and some hints can help an operator accomplish this task without damage to the utility.

6.1.2 Preventing Accidents

One of the best preventive measures an operator can take toward safely locating utilities is to be familiar with the area where the excavation is to take place. At some point, underground utilities terminate, surface, incorporate a valve into the system, or are sectioned by a pull box or vault. All will result in some type of visible surface indication.

> *Some types of flagging ribbon contain a thin metal strip that allows the ribbon to double as a locating wire or tracer. If this ribbon is damaged, it must be repaired.*

On the first day at a new location, an operator should arrive a few minutes early to take a walk around the working area to look for indications of underground utilities. For example, an operator should look for valves in the street. The valve alone will indicate that there is an underground line; two like valves in the same line can indicate the direction of the line. Utility poles carry telephone, electric, and cable television lines overhead. If there are no lines running overhead, the operator should not assume there are none in the area, because they may be running underground.

If work is being done around a traffic intersection, the operator should look for pull boxes on each corner. There will be a conduit lat-

Underground Utilities

W	Water
G	Gas
S	Sewer
D	Drain
E	Electric
T	Telephone
O	Oil
GS	Gasoline
TV	Television

Typical utility marks.

eral extending from the begin curb return (BCR) to the end curb return (ECR) on each corner, as well as a conduit crossing from one side of the street to the other. On one corner of the intersection will be a metal cabinet called a controller. This controller is the brain of the intersection, and in some cases it may have as many as six underground conduits feeding it. One of these conduits is the main power feed, and it is one of the few conduits that will run away from the intersection. Even when work is being done a considerable distance from an intersection, it is still possible to make contact with this conduit.

> **SAFETY TIP**
> An operator should not dig in an area that has not been surveyed by a utility company.

Digging behind a curb in a sidewalk area close to an intersection should be done carefully because of the conduit that is carrying the power to the intersection and because traffic signals use a small-gauge, telephone-type cable to coordinate signal timing. This cable is called interconnect and is usually laid in a conduit within a couple of feet of the back of the curb.

Street lights are another good above-ground indicator that something lies below. If there is more than one streetlight in a line, a conduit is probably connecting them.

6.1.3 Utility Location

Underground Service Alert and Digalert are two types of utility location services available to mark the location of underground utilities.

If the utility company has not contracted one of these independent location companies to mark its utilities, it is the utility company's responsibility, upon proper notification, to ensure correct marking of its utilities so that safe excavation of the area can occur. When a utility location has been marked, the area surrounding the mark down to the actual utility should be pot-holed by hand. However, there are occasions when, due to location, depth, or type of material the utility is buried in, a backhoe loader will be needed for partial excavation or

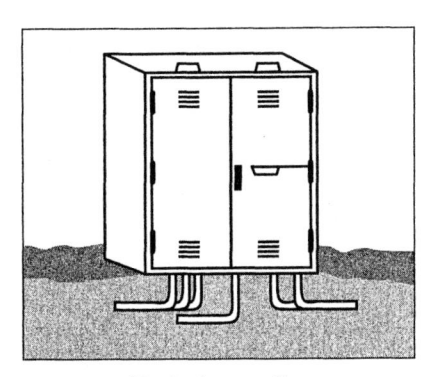

Typical controller.

Chapter Six: Utilities

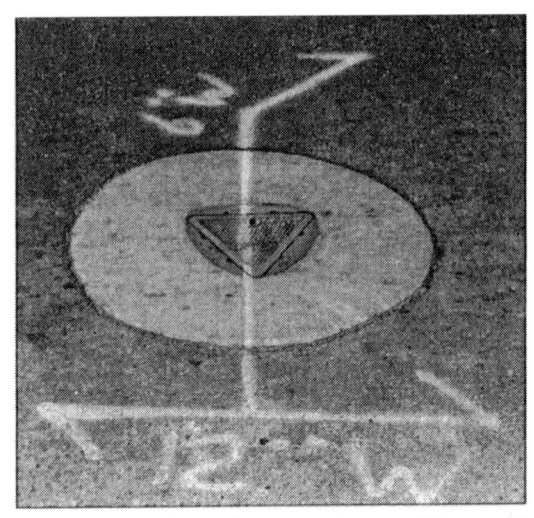

An operator should pay close attention to utility markings.

actual location of the utility. The marks painted on the ground that indicate the location of the utility will have a minimum variance of 2 feet on either side so the operator is responsible for the utility for 2 feet on each side of the utility mark.

6.1.4 Recognizing Trench Signs

An operator doesn't always have to depend on tooth contact to find the utility as certain clues will present themselves before the teeth make actual contact. For example, caution tape, also called flagging ribbon, is a thin sheet of plastic about 3 inches wide and about 1,000 feet in length. This tape is placed above the conduit or pipe in the trench immediately before backfilling so that as the operator nears the utility with the bucket, the banner will be spotted in the spoils, letting the operator know the bucket is within range of the utility.

Once the tape has been spotted in a trench or in the spoils, all backhoe operation should cease and digging should continue by hand until the utility is found. Operators need to keep in mind that one can never be sure where the tape lies in relationship to the utility in the trench. Sometimes, for example, the tape will be taped directly to the conduit it is supposed to be protecting.

6.1.5 Watching the Spoils

Operators should always be watching the spoils, whether or not a utility is being located. Trench spoils can tell the operator a great deal about the material being excavated. For example, the material in a trench that has been backfilled will always be slightly different from the surrounding material. Even when the backfilled trench contains the same material that it was originally dug from, the material will lay back in the trench differently from how it was laid before it was disturbed. An operator keeping a sharp eye on the spoils will notice a change in the material as a trench line is crossed. Once a trench line has been noticed, careful digging will reveal the buried line below.

> **SAFETY TIP**
> Abandoned utilities may still contain flammable gases. Whenever possible, an operator should try to work around the utility instead of going through it.

Certain utilities have different backfill and compaction requirements. For example, an irrigation line in the middle of a field isn't going to require the same compaction as a jet fuel line buried beneath a freeway. Some types of utilities require sand under and over the utility. Depending on the stability of the surrounding material and the type of utility, one- or two-sack slurry may be required instead of, or in addition to, the backfill material.

6.1.6 Location Technique

Operators feel the density in material from one pass to the next. If an operator sinks the bucket's teeth into the ground and the teeth grab a conduit, the conduit will break in half. When the teeth sank into the ground and the operator first started to pull in, the first feel was that of the conduit. As the operator gradually pulled in farther, the conduit was being slowly torn in half. The operator probably would not have felt the change until the conduit snapped.

Operators can avoid this by always making sure that the first few inches of the bucket pass are in dirt. When an operator is beginning a bucket pass in the trench, the teeth should not be abruptly sunk into the ground. Instead, as the bucket is being pulled toward the operator from the start of the trench, the teeth should slowly penetrate the dirt and enter the trench while the bucket is moving. As the bucket nears the middle of the trench, the operator should slowly increase pressure on the crowd arm, thereby increasing the depth of the cut.

> *Movement of the surrounding dirt is often the first indicator that an operator is on a utility.*

As the operator nears the end of the trench, the bucket should be slowly boomed up, decreasing the depth of the cut. The bucket should not be curled at the end of the pass. This will only increase the chances of hooking an unseen utility.

After a few passes the operator will see the profile of a bowl starting to take shape in the trench. If the operator is accustomed to digging a square, flat trench, it may be uncomfortable to dig a bowl-shaped trench. After the utility has been safely located, the trench can be squared up.

Another way to avoid breaking an unseen utility is to be sure the first couple of passes are dug no more than 3 or 4 inches deep. The

Chapter Six: Utilities

An operator will see the profile of a bowl taking shape in the trench.

next pass should be kept within the limits of the first pass, but the operator should shave 2 or 3 inches off one side of the trench or the other. This extra couple of inches in width needs to be continued until the utility is located.

6.1.7 X-pattern Digging

When entering the trench for a pass, the operator will need to stay tight to one side and, as the teeth start to penetrate the material at the depth of the pass, gradually shift the bucket (still at the same depth) to the opposite side of the trench. On the next pass, this pattern should be repeated except in reverse, making an X in the trench. On the next two passes, the operator should pull in straight on both sides, giving an even grade on the bottom of the trench.

> *Too little r.p.m. hampers the hydraulic flow. Too much r.p.m. hampers the response time when contact is made.*

The reason for this criss-cross pattern is that, as the operator is slowly crossing from left to right, the teeth are making contact with a conduit running parallel in the trench. The bucket will ride against the side of the conduit, disrupting the pattern. This disruption can be easily noticed by the operator by watching the bucket carefully as a pass is made.

If, when an operator is pulling the bucket in straight and making this criss-cross pattern, the bucket starts to move in any direction other than what is intended, the operator knows that a conduit has either split the teeth or is riding alongside the bucket. This extra trench width will also allow the conduit to give a little in case teeth contact is made. When digging slowly, this give should supply ample back pressure through the controls to allow the operator to back off before damaging the conduit.

6.1.8 X-pattern Variation

Another variation on the X-pattern location technique is used when it does not matter how much material, which is surrounding the utility location mark, is removed. The operator should position the backhoe so that most of the digging will occur in the sensitivity zone. With the utility location mark as the centerline of the dig, the operator should start with the first pass 6 or 8 inches in depth by approximately 4 feet long. Instead of moving over just a couple of inches in the X-pattern, the operator should take a full bucket width pass next to the one just taken, again at 6 or 8 inches in depth. With the second pass, the edge of the bucket should have produced spoils that flowed into the area of the first bucket pass; the operator should make a quick pass to clean this. The third pass should be on the opposite side of the trench from where the last pass was made, widening that side of the trench by a bucket width. Another pass should be taken on the spoils that were made in the middle by the previous pass. In the end, the operator should have a nice clean area, about 4 feet square and 6 or 8 inches deep, with the utility location mark at the center.

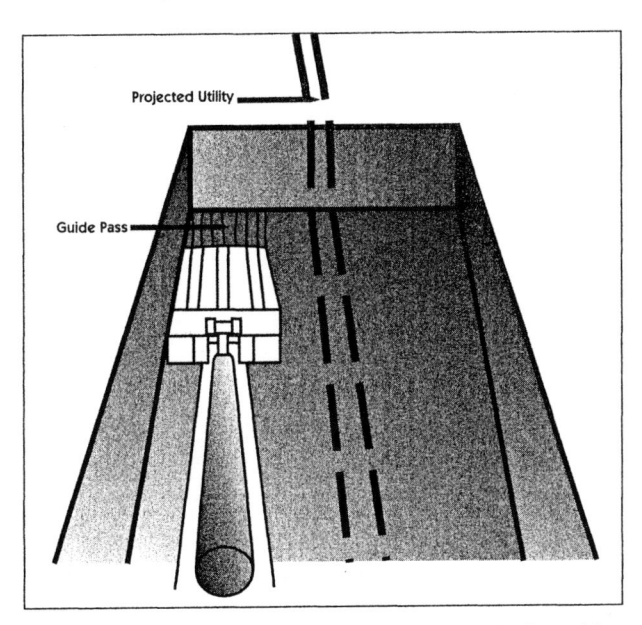

An operator should keep all tooth penetration in the guide pass area, moving diagonally as the bucket is drawn in.

Chapter Six: Utilities

Now an operator must dig as if the utility will be found on each pass. In order to locate the utility safely, the operator should follow two rules. First, a guide pass should always start the section drop and is the only stroke where the teeth should enter the material vertically. Second, the operator should not change the angle of the backhoe or the bucket until the utility is located.

The first pass should take place on either side of the area, as far away from the utility mark as possible. This is the guide pass. This pass is not intended to find a utility but to give the operator a safe place to sink the teeth and give the operator a start into the material.

The guide pass should be made along the entire length of one side of the area, no more than 3 inches deep. The operator should pull the material toward the machine and leave it in a pile as close to it as possible.

The next pass should start at the top of the first pass and run diagonally to the opposite side closest to the operator. This pass should be made slowly and at the same depth as the guide pass.

After this pass is complete, the operator may determine that the utility is not in the area at that depth. In that case, the operator should clean the corners and the rest of the material from the area, making sure not to dig the teeth any deeper into the grade than for the guide pass. The operator should start at the side from where the first guide pass was made and another should be started. The same digging sequence should be followed until the utility is safely located.

6.1.9 Parallel Setup

When the utility mark indicates the direction of the utility, the operator needs to decide whether to set up parallel to or perpendicular to the utility. Both have their advantages and disadvantages.

There are a number of disadvantages in setting up parallel to the utility. First, when plastic is involved, the operator runs the risk of slicing the conduit with the teeth. The operator also increases the possibility of snagging a joint or a coupling, and an operator can disfigure or even break a conduit that is not running perfectly parallel in the trench.

Second, from the seated position behind the bucket, an operator

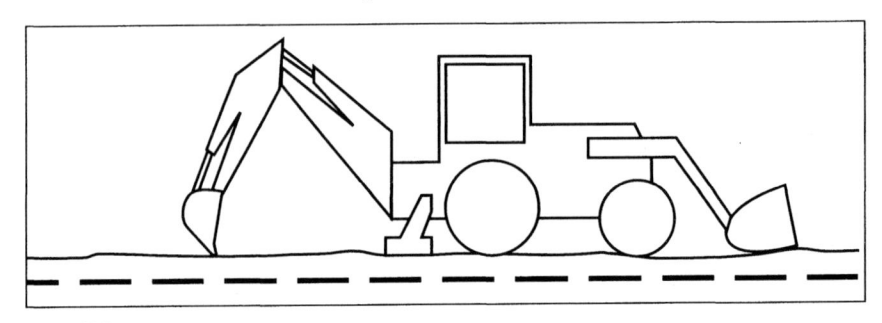

Parallel setup.

Locating Utilities

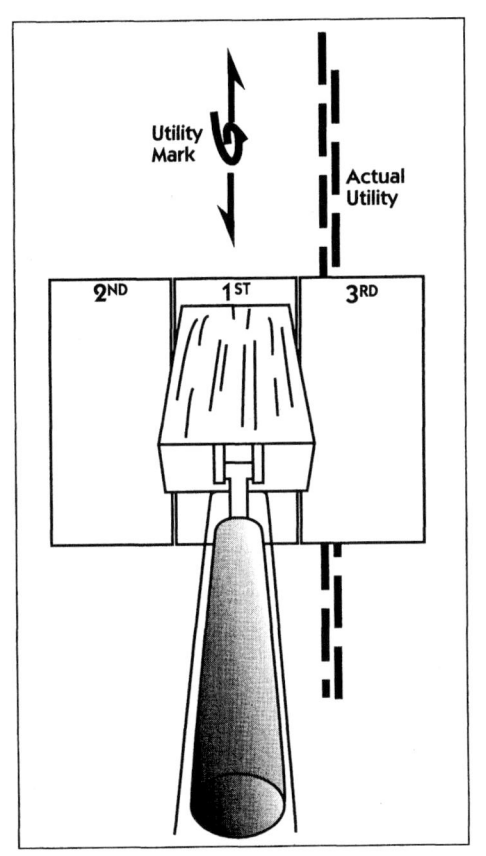

One drawback to a parallel setup is the potential difference between the projected utility versus the actual utility.

cannot see if he or she has uncovered or made contact with a conduit for 3 to 4 feet when the bucket is far enough in and the operator can see over the top of it. For these reasons, it is imperative that the operator has good help on the ground.

Third, since utility marks are seldom correct, there is a good chance that the operator can center the bucket directly on the utility mark and still not find the utility on the first dig. In this case, the operator will need to move over to one side of the mark and continue widening the original hole until the utility is located. Consequently, parallel locating is generally slower than setting up perpendicular.

> *The operator should take the same amount of material on each bucket pass, while keeping the bucket moving at the same speed.*

6.1.10 Perpendicular Setup

Setting up perpendicular to the utility is the quickest and the most dangerous, because contact is made with the teeth or the movement

Chapter Six: Utilities

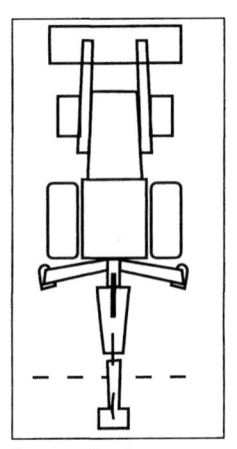

Perpendicular setup.

of the surrounding material at the instant contact is made. The success of this method relies on the operator's understanding of what areas can and cannot be felt with the bucket.

6.1.11 Sensitivity Zone

The area of greatest sensitivity for a backhoe is in the second and third quarter of the backhoe's overall travel distance.

An easy way to find this is for the operator to reach out as far as possible with the bucket, then pull back in a couple of feet and mentally mark this point. The operator should then bring the bucket in as far as it will come, push it back out a couple of feet, and mentally mark this point also. The backhoe should be positioned to where the utility mark falls between these two points. This is the sensitivity zone.

The beginning of the first pass should start at the first mental mark the operator made, and it should end at the second mark. All locating should be done inside this perimeter. Should it become necessary to dig closer or farther, the operator should move the backhoe the necessary distance while still keeping the excavation in the second and third quarter. It's a good idea for an operator to practice this technique, developing the feel before trying it on a job.

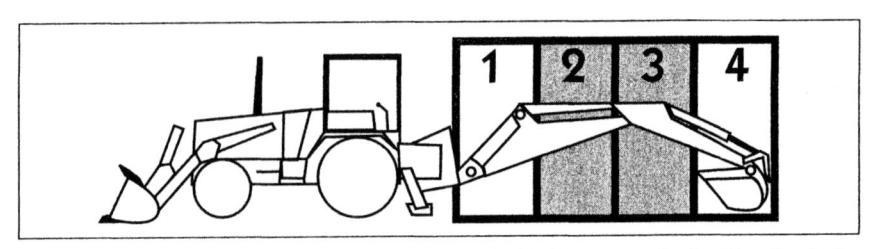

The sensitivity zone is located in the second and third quarter of a backhoe loader's overall reach.

Locating Utilities

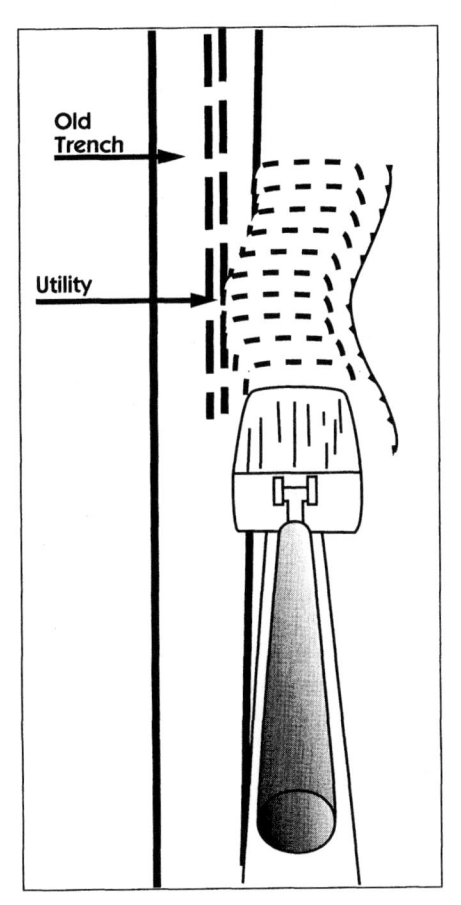

If the material is soft, the operator can reduce the possibility of damage to the conduit by using the outriggers to change the angle of the teeth slightly away from the conduit.

6.1.12 Looking for Existing Conduit

An operator has laid the conduit in the trench, gone to lunch, and returned to find that part of the trench has been backfilled by a skip loader making a path for a truck to unload material on the job. Because this was not expected, the operator did not bother to mark the location of the end of the conduit. But now the operator needs to extend the conduit run another 300 feet.

To find the end of the conduit, the operator should set up parallel so that the trench line is off-center to the backhoe by a foot on either side. Since the operator can't see the trench, good judgment must be used to dig beside the existing trench. If there are no utilities in the way, and the operator is confident that the digging is being done next to the old trench and not on top of it, then digging directly down to the same depth as the old trench is fine.

Chapter Six: Utilities

The new trench should be at least 4 or 5 feet long. Once the same depth is reached, the operator should start by lowering the teeth to the bottom of the grade and shaving the side closest to the old trench. The shaving should be done in a gradual motion, starting easy against the opposing trench wall and increasing in swing pressure until the bucket is half the distance of the new trench. The operator should then slowly decrease in pressure until the operator is no longer shaving into the trench wall.

It is very important that every time the operator makes a new shaving pass, it is started with the teeth at the bottom of the trench. Instead of running the risk of tooth damage to the conduit by locating it from the top working down, the operator is using the side of the bucket to shave away the material until it can be visually seen, or until the operator can feel the change in pressure against the side of the bucket.

6.2 Locating Electrical Conduit

The biggest advantage an operator can have when locating electrical conduit is to rely on the eyes, not the bucket's teeth. If an operator is locating a utility that has been marked by a utility company, then the operator will have a general idea as to the utility's location.

When it comes to electrical lines, an operator should not depend on any one person or method to mark the utility location. Whenever possible, an operator should back up a utility mark with a personal evaluation. This evaluation should be based on existing above-ground electrical structures.

6.2.1 Types of Conduit

Electrical cable is carried to its source in three ways: in conduit, either plastic or metallic; in cable in conduit (CIC), which is a very thin tubular sheathing that comes in roll form; or in the form of direct burial where the cable is laid directly in the grade without any protection.

> **SAFETY TIP**
> Ground acids can eat the mastic coating from older gas lines, making them look like electrical conduit.

Before determining the backhoe's setup position for the electrical line to be located, it is helpful to find out what type of conduit is being used. Although an operator can't see the conduit through the ground, there are certain clues one can use to ascertain the type of conduit.

6.2.2 Examining Pull Boxes and Vaults

Pulling wire long distances through conduit is difficult. To make the job easier, the entire distance of the wire pull is divided into equal sections. Depending upon the gauge of the wire being pulled, each section is divided by a pull box or a vault. In either case, the conduit en-

An operator should never stick a metallic object in a pull box or vault.

ters through one side of the structure and exits through the opposite side, leaving a void between the two ends to pull the wire through.

Lifting the lid of a pull box closest to the area in which the operator is working will allow inspection of the conduits to determine their composition. Lifting the lid on a pull box not only shows the operator the type of conduit being used, but also the depth of the conduit and the direction in which the conduit is laid.

The depth can be checked by taking a long, thin nonmetallic object smaller than the inside diameter of the conduit (a grade lath works well) and pushing it down inside the conduit until it makes contact with the bottom, being careful not to nick any wires. By measuring the rod or lath, the operator can get a pretty good idea of the depth.

Often the utility marks are headed toward an electrical vault transformer. Underground electrical vaults are indicated either by manhole covers set directly on the vault or by the transformer sitting on a concrete pad above ground level. In either case, the electrical cable will be carried in plastic conduit into the vault.

Underground vaults often contain flammable gases. An operator should never remove the lid from a vault without notifying the utility company first. The company will either require a member of their field personnel to supervise the removal of the lid or the company will remove it.

> **SAFETY TIP**
> Underground electrical vaults often contain flammable gases. An operator should never remove the lid from a vault before notifying the utility company first.

6.2.3 Utility Poles

Telephone or utility poles will often be in line with the utility mark. If the utility is drawing power from, or terminating at the top of, the pole, the operator will see the conduit coming up from the ground and running up the pole.

If an operator is digging next to a utility pole, the electrical company usually requires a sweep going up the pole. Sweeps in a trench start the transition up the pole 3 or 4 feet back in the trench. This means

Chapter Six: Utilities

that if a conduit starts the transition 4 feet away from the pole at 30 inches deep, it raises to 15 inches at 2 feet away from the pole. This needs to be kept in mind when digging close to a utility pole.

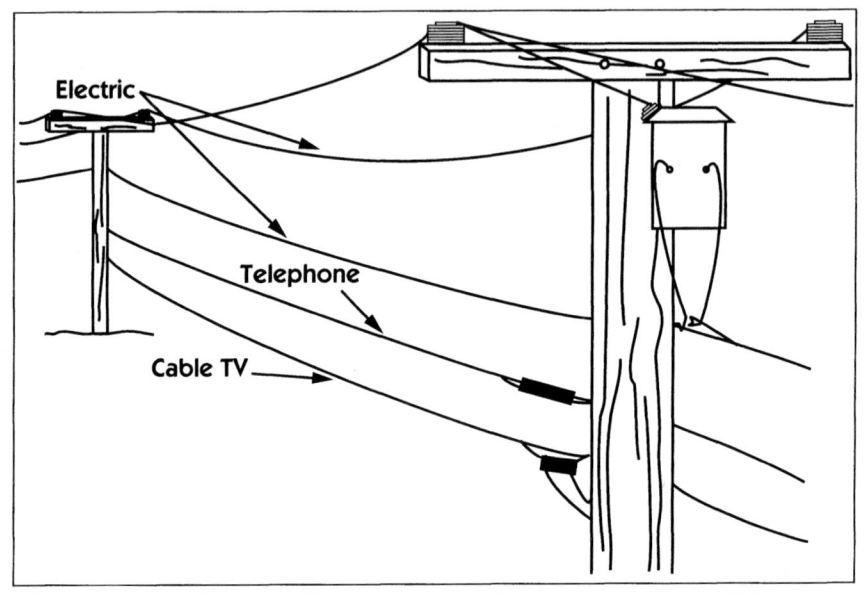

Utility poles usually carry utilities in a pattern: high voltage on top, telephone in the middle, and cable television at the bottom.

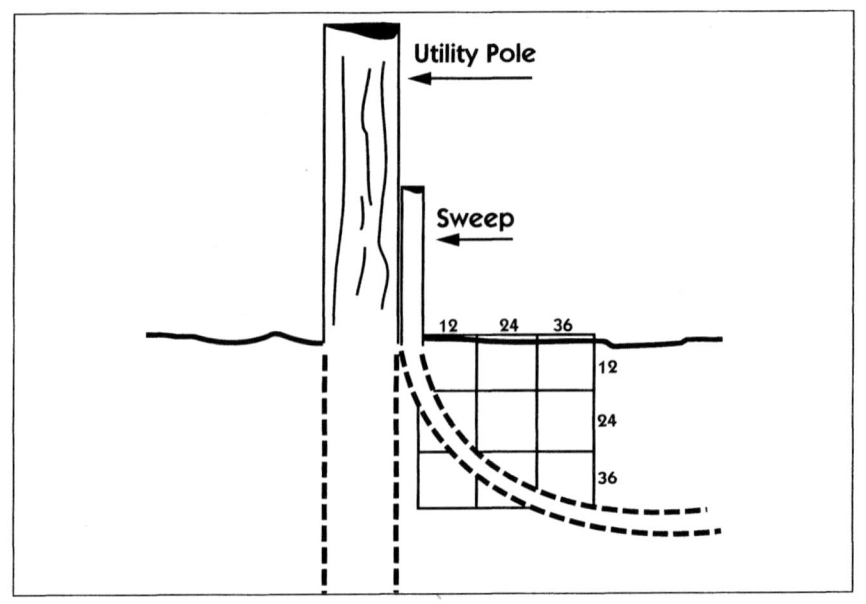

Sweeps should cause an operator concern when digging next to a utility pole.

6.2.4 Setup

Once the operator has gathered all of the information possible to help ascertain the type of conduit being used, it's time to set up.

If the conduit is metallic, setting up perpendicular is the best option. Although metal conduit will take quite a bit of abuse, care should still be exercised because metal conduit will break when enough force is applied.

Setting up parallel to locate a conduit safely requires help on the ground. With a laborer's help, setting up parallel to the conduit is best when there is a question as to the type of conduit or if the conduit is plastic.

When an operator is locating a conduit alone, one will need to rely on tooth contact to locate it. The operator should set up perpendicular, taking an inch of material at a time and not forgetting to dig the trench a couple of inches wider than the bucket to provide slack when contact is made.

6.3 Locating Water Lines

The size and type of conduit used to carry water underground depends on the water's volume and pressure. There are no general guidelines as to where and when a specific type or size of conduit is to be used. This is determined by construction codes and application purposes.

As with most utilities, an operator cannot depend on sand or slurry to warn that a water line is buried below. An operator can hope for a soft spot as an existing trench is crossed, or the operator might see the different backfill material at the cross section in the trench, also indicating a bisecting trench.

But the operator is the only one who knows what it feels like when the

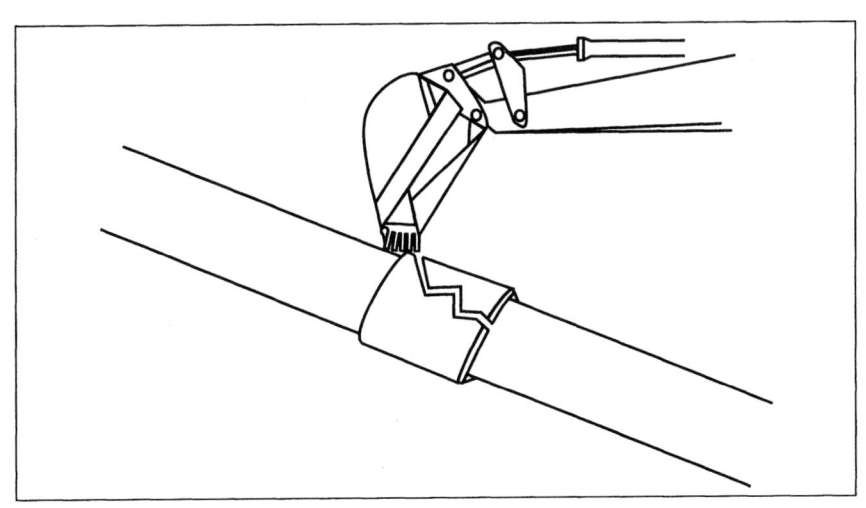

Transit couplings under pressure can easily crack when struck with a tooth.

Chapter Six: Utilities

> *If a water line is buried in native material, an operator needs to rely on sense of feel to locate the line.*

teeth roll over the top of a transite line, who can feel the 2-inch main line buried 8 inches under the ground starting to give under the pressure of the teeth, and who to blame when the ¾-inch line, the one the operator knew would not be felt in the rocky material, is torn apart.

Understanding the different types of conduit that water is carried in will allow the operator to make the decision about which digging technique to use to locate the line safely, and how much abuse the line will take before damage is inflicted.

6.3.1 Transite Concrete

Transite is also known as asbestosized concrete (AC). Although it is slowly being phased out in favor of plastic conduit, there is still a great deal of transite pipe in place. Transite is a very strong conduit that can withstand a great deal of internal pressure. It is also brittle and extremely susceptible to shattering or cracking when struck with one bucket tooth. Transite pipe is joined by couplings to complete a section. These couplings can be either fixed on one end of the transite length, allowing another length to be slipped into place, or the coupling can be used as an individual piece, able to be slipped on the end of any piece cut to size. In either case, the coupling is secured to one or both ends by a rubber grommet that is slipped into a preformed

Transit coupling with preformed groove for rubber seal.

groove on the inside of the coupling.

Because the lengths of pipe are held in place by this rubber grommet, and not glue or mortar, it makes the sections vulnerable to movement at the coupling. When digging next to or crossing a transite water line, an operator should pay extra attention to the location of these couplings. An operator needs to be careful not to catch a coupling on the side of the bucket because it will move the coupling and possibly crack it. A water line making a gradual turn, such as one following the radius of a street, needs to be given extra care because there is a great deal of pressure being exerted on the couplings. The backfill material around the pipe is the main component in keeping the line intact when it is under pressure. Operators need to be very careful when removing this backfill material from around a water line on a radius.

> *Transit couplings under pressure can easily crack when struck with a bucket tooth.*

6.3.2 Plastic Conduit

The use of plastic conduit for water lines is very widespread and is gaining popularity as stronger and more versatile plastics are developed. Plastic conduit is light, very strong, and can be cut to any length and recoupled quickly.

Glue is used to bond smaller diameter conduits that have preformed bell ends. On conduits that have a diameter of 4 inches or more, a pressure-sensitive tapered grommet, similar to the type used on transite pipe, is used to seal the conduit. This grommet seal method is a very effective means of sealing the conduit, especially when it has been pressurized with water.

> *As with any segmented conduit, the vulnerable points are the couplings and the outside edge on a radius.*

Plastic water lines are made in all sizes.

Chapter Six: Utilities

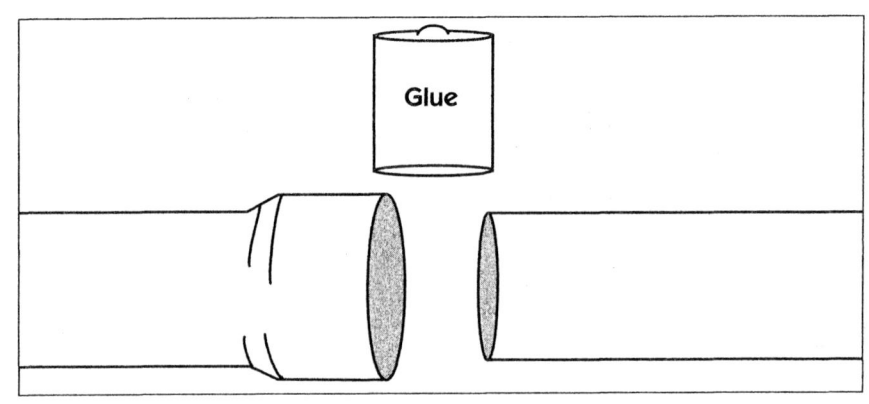

Small plastic water lines are sealed with glue.

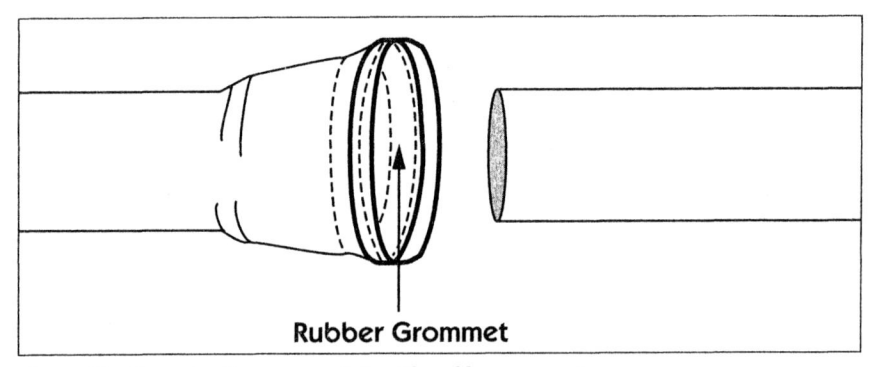

Large plastic water lines are sealed with rubber grommets.

As with any segmented conduit, the vulnerable points are the couplings and the outside edge of the line on a radius. Catching a tooth on a coupling while a pressurized line is exposed on a radius could result in disaster.

6.3.3 Steel Conduit

Ductile iron is the most commonly used steel water line, and it is used in a number of different forms. It can come wrapped in a thin layer of concrete, except for the ends that are left exposed to allow joining of the lengths, or it can come wrapped in a mastic coating similar to the coating used by the gas company on their conduits. Other ductile iron lines may be covered with a plastic slipcover that is slid over the pipe during assembly. All of these coverings are used to protect the conduit from corrosion.

If the wrap or coating on a steel water line becomes damaged, it must be repaired to keep the line from corroding.

Locating a steel water line is generally not a problem. If the line is sanded, then the sand will be the operator's indicator. If the line is laid in native material,

It takes quite a bit of effort to break a ductile steel water line.

the operator will need to rely on the sense of feel to find it.

If an operator is digging at a moderate pace, the only damage that will occur upon tooth contact will be a chip in the concrete or mastic covering, or a tear in the plastic slipcover.

Although steel water lines have been known to break, it requires a great deal of effort on the operator's part to do so.

6.4 Locating Gas Lines

Gas lines made with the plastic material polyethylene (P.E.) are easily identified by their color, ranging from yellow to orange. If the line has been in the ground for a period of time, it will turn gray, similar in color to plastic conduit. If the situation ever arises where an operator needs to notch a conduit to identify its contents, care should be taken because it may be an old live plastic gas line.

> **SAFETY TIP**
> Yellow or orange plastic gas lines can turn gray, making them look like electrical conduit.

If the gas mark doesn't specify P.E., it is a relatively safe assumption that the gas line is made of steel. Knowing the composition of the gas line will help the operator determine the position needed to locate the line.

6.4.1 Sand as an Indicator

Backfill sand is a good indicator that a gas line is in the area, especially if the sand coincides with a utility mark. But assuming that digging can continue until sand is hit when looking for a gas line is to set up for disaster. A good example of this is the fact that the gas company, or the subcontractor hired to do the job, may have chosen to bore the gas line instead of tearing up a street with trenching.

Chapter Six: Utilities

This gas line is made of 1-inch diameter steel in one section and ½-inch diameter plastic in another.

6.4.2 Boring

Boring is a technique used to get a conduit from point A to point B without having to trench or disturb the surface material. A jacking pit is dug and, with the help of a steel slide rail and a hydraulic boring machine, a steel conduit is bored under the ground to the desired point. The steel conduit is then pulled out and the plastic or steel gas line is pushed into the hole left by the previous conduit.

Boring is especially useful under freeways or heavily used roadways where stopping or diverting traffic is not feasible. In addition, sanding a gas line that has been bored is impossible.

6.4.3 Setup

Setting up to a gas line for location should be done at a perpendicular position, with the operator making sure that the contact point will be in the sensitivity zone. Setting up parallel to the line increases the possibility of slicing the line with one of the teeth.

If the gas line is known to be P.E., an operator should follow the advice in Section 6.1, "Locating Utilities," and dig the location hole a few inches wider than the bucket. Plastic gas lines can stretch quite a bit, and the few extra inches will be beneficial when teeth contact is made.

An operator should not forget about the tracer wire that often accompanies a plastic gas line.

After successfully locating the gas line, the operator should make sure, before starting to push the material away from the line to continue the work, that the operator finds the small-gauge locating wire, or tracer wire, that normally accompanies a gas line. This tracer wire is used to locate the line in case of a gas leak or when future excavation becomes necessary.

6.5 Locating Telephone Lines

Telephone line is easy to break and expensive to repair. When the conduit is encapsulated, the conduit can be anything from redwood box to transite to clay pipe to plastic to direct burial. Telephone lines can be buried in native material without any sand shading, backfilled with full sand, or encased in slurry. The depth of a telephone line can range from 18 inches to 5 feet, and the conduits can be assembled in banks of 20 or more.

> *It's **not** okay to dig.*

Because of all the variables involved with telephone line, an operator must exercise extreme caution and, most important of all, **know when not to dig.**

There are a couple of tips that can help an operator reduce the possibility of damage. First, an operator should always have a competent laborer. It is this person's job to see what the operator cannot, and very often this can mean the difference between locating a line and tearing one in half.

Second, the operator should make sure the contact area is in the sensitivity zone, and that the location hole is large enough so that the front or back wall of the hole will not come in contact with the bucket while locating the line. It is difficult to concentrate on finding a utility when the back of the bucket keeps rubbing on the back wall each time a bite is taken.

6.5.1 Fiber Optic

Fiber optic is telephone's replacement in the near future. With fiber optic, the information normally carried in electrical impulses on a conventional telephone wire is transmitted by light on very fine glass threads.

If an operator breaks a live fiber optic line with the backhoe, the operator will probably lose his or her job. Chances are also very good that the equipment being used will also be lost. An operator should never get the teeth anywhere near fiber optic. If a line must be exposed, an operator should do it by hand.

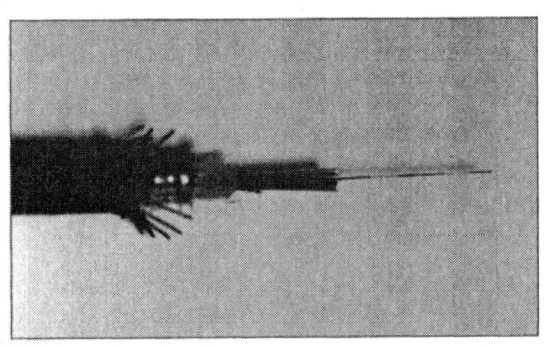

This stripped fiber optic cable reveals the outcasing protecting the inner glass fibers.

6.6 Making Contact with Utilities

Unfortunately, breaking utilities is an inevitable part of digging in the ground. No matter how good an operator's sense of feel is, a utility will occasionally break. If an operator is trying to locate 30 one-inch plastic water lines, the odds are good that one is going to break.

Making a mistake is one thing; compounding it by not correcting it is another. If a utility is hit, an operator should not stand around wondering what went wrong. The operator needs to act immediately to solve the problem. For example, if a gas line is ruptured during the course of digging, the gas company should be called immediately to inspect and repair the damage, because an operator in the field does not have the available tools or knowledge to repair the ruptured line. But what should an operator do when, upon close inspection of a traffic signal conduit that has been made contact with, the conduit is cracked but the wires inside are all right? Backfill it and forget about it? No, that would be wrong. The operator could call an electrical contractor out to repair the line properly, but time would be lost.

An operator should not make a mistake worse by not doing something to correct it.

The other option is for the operator to fix it **properly**. As long as the operator has the knowledge and the material to do it right, then it should be done. At the moment of break, all responsibility rests on the operator's shoulders. It is a field judgment call.

An operator needs to keep two things in mind. First, poor workmanship will always come back to haunt an operator. Second, if an operator has **any** reservations about the work that needs to be done, the utility company should do the work. It is well worth the money to have peace of mind.

6.6.1 Making Contact with Electrical Conduit

Electricity should never be taken for granted. Too many accidents have occurred when an operator bit into a live conduit that was presumed deactivated or abandoned; electrical conduit should always be treated as though it was alive.

Operators should be careful around all electrical conduit.

When teeth have made contact with an electrical conduit, an operator's decision on what to do next is based on a number of variables. The more an operator understands about electricity and its transmission, the better equipped he or she will be to handle adverse situations.

6.6.1.1 Electricity Transmission

Electricity is generated at a main plant and sent to substations by high voltage transmission lines. These transmission lines, sometimes

Making Contact with Utilities

High voltage carried overhead.

referred to as trunk lines, are either overhead on utility poles or underground in vaults, depending on the voltage being carried and the distance between substations.

From the main transmission lines, the electricity is carried to transformers. These transformers change the raw high voltage electricity into a usable voltage. The reduced electricity is then carried to the electrical panel of the structure that is being fed.

The smaller, reduced electrical lines coming out of the transformer and heading toward the building or the electrical panel are generally 660 volts or less. If an operator breaks one of these lines with the teeth, anything from a small smolder to a brilliant flash can occur. Although accidents around these smaller arterial lines are rarely fatal, everyone digging around them, including the laborer, should remain a safe distance away.

6.6.1.2 Breaking a Line

When an arterial line is broken, the surge of electricity going into the ground will overload the transformer and trigger an internal breaker which stops the flow of electricity to the broken line. However, when a break occurs in a main feed line (generally 12 kv or more; kv = kilovots; 1 kv = 1,000 volts), other things can happen.

When the break in the main line causes the electricity to ground, the surge of electricity can blow up the smaller transformer and con-

Chapter Six: Utilities

tinue to destroy transformers in the electrical flow line until the surge of electricity has been diminished enough for a transformer in the line to stop the flow. All this takes place in just a few seconds.

In some cases, if the electrical surge is so great that the transformer can't hold the electricity, a signal is actuated at the substation indicating a problem in the line, and the circuit is automatically shut down. Whoever receives the signal at the substation knows there is a problem in the line because the circuit has automatically shut down. However, this person doesn't know if the problem is in the overhead transmission line or underground in a conduit.

> **SAFETY TIP**
> When contact has been made with a main electrical feed line, an operator should stay on the backhoe loader until the operator is certain the circuit has been closed.

When an operator has made contact with an underground high-voltage line, the operator will know it immediately. Depending on how the teeth penetrate the wires in the conduit and how many of the larger phase wires are opened, causing a cross short in the conduit, an operator can expect anything from a flash with a loud buzzing sound to a small explosion that takes part of the bucket with it.

When the smoke clears, an operator's first reaction may be to pull the bucket up out of the trench or to jump down off the backhoe to see what the teeth are in. Operators should resist these urges. With high voltage breaks, someone at the substation will probably try to clear the line by sending a surge of electricity through it. Consequently, after the initial break, the operator can expect two or more electrical surges at about 30-second intervals. If the teeth are still in contact with the broken line when the surging occurs, your backhoe will act as the closest grounding point and so will take the brunt of the electrical charge.

Electricity that is in the process of finding ground can, and usually will, arc. Anyone standing in the vicinity of the break in the line or of the backhoe runs the risk of electrocution from these arcs. An operator stepping off the backhoe while the backhoe is grounding the short will be electrocuted.

Should an operator's teeth penetrate a high-voltage line, everyone should be kept away from the area and the operator should stay on the backhoe until everyone is absolutely sure the circuit has been closed and any risk of electrocution has passed.

6.6.2 Making Contact with Water

> *An operator should find the shutoff valves in the area before digging.*

Unless a water line is very large, breaking it is generally more of a nuisance than it is dangerous. Nevertheless, if not taken care of immediately, a break in a water line can do extensive damage to a new construction zone. If water is suspected to be in the area, it is always a good idea for an operator to take a minute before digging to find out where the shutoff valves for the lines are.

Water that is not under pressure, such as puddling water from a broken line, will seek the lowest point. When a break occurs, it is the operator's job to make sure the lowest point is in a catch basin or other safe place where the water can't cause damage.

An operator should always pay close attention to curb-and-gutter, sidewalk, and street grades in new construction areas. These areas are easily ruined by excessive water and they are expensive to regrade, especially when the subgrade becomes saturated.

> **SAFETY TIP**
> When a large line break occurs, an operator must act quickly. Water under high pressure can collapse a road in a matter of minutes.

The natural digging motion of the backhoe means that when a line is broken, the water will either shoot up into the air, making observation of anything difficult, or it will shoot back into whatever the operator is digging. If the break is close to the backhoe, the operator should lay the bucket flat over the break to divert the water off the bottom of the bucket. If the break is farther out, the operator should open the bucket all the way and rest it directly over the break, diverting the water back to itself.

Once the water is spraying in a direction where the operator can get close to the break, the operator will need to shut off the valve to stop the flow completely, or get a hand in the hole created by the spraying water and determine the size of the conduit and what it's made of. If the conduit is plastic or steel, nothing can be done to stop the leak without shutting off the valve. In this case the operator needs to concentrate on diverting the water to save the grade, even if it means removing the bucket from its resting place on top of the leak so that a drainage trench can be cut. If the broken line is made of copper, it can be crimped and stopped at the break.

6.6.2.1 Crimping Copper

Crimping a copper line to stop a leak in the field is a judgment call. If the operator knows where the shutoff valve for the line is located, it should be shut off. But if water is threatening to cause damage to the grade, the line needs to be crimped.

The operator should start by lifting the bucket from its previous position over the break and note the water stream direction. The operator should follow the stream of water to the conduit with the teeth. When the teeth make contact with the conduit, they should slide along the top or back of the conduit until they are just far enough from the break so that the operator can curl the bucket while holding the conduit. By hooking the teeth under-

> *If the broken line is an irrigation line, an operator should be careful not to damage the control wires that often run below the line.*

Chapter Six: Utilities

neath the conduit and pulling it straight, the operator will be able to see the precise direction the conduit is running.

If the conduit is running perpendicular to the position of the backhoe, the operator should place the side of the bucket against the broken end of the conduit that was just pulled up and swing sharply toward the direction of the conduit, folding it upon itself. Because the operator is trying to kink the conduit and not bend it, the operator may need to boom down slightly if the bucket starts to ride up the conduit. If the conduit is running parallel to the backhoe, the operator should hook the teeth under the end of the conduit and pull up slightly and straight back.

> **SAFETY TIP**
> An operator should wear gloves when working on conduit that is submerged in water because sharp copper or plastic edges can cut a hand.

If the water is coming from the opposite end of the conduit, the operator should place the back of the bucket against the end of the conduit, applying enough pressure to hold the conduit, and then boom up slightly and push straight back. If the flow of water has slowed down, a bend in the conduit has been established. If the operator pushes the conduit farther upon itself, the flow of water should either stop or be light enough so that the job can be finished with a hammer.

6.6.3 Making Contact with Gas Lines

An operator will know immediately when the teeth have penetrated a gas line. The chain of events that follow a rupture are the same whether the gas line is plastic or wrapped steel.

Gas contained in its conduit in a pure state is very stable and is generally not a problem. Pure gas, generally, is not flammable without a direct flame. The problem arises when the line is ruptured and the gas is allowed to enter the atmosphere and mix with surrounding air. Even then gas only becomes flammable when the proper gas-to-air ratio occurs.

It is natural to assume that a rupture in a high-pressure gas line is more dangerous than a rupture in a low-pressure line because of the enormous volume of gas that escapes from a rupture in a high-pressure line. But while any gas line, high-pressure or low-pressure, is very dangerous when ruptured, a low-pressure line tends to be more dangerous. Because of the lower volume of gas escaping, the gas is able to mix with the air and find its optimum flammable ratio much quicker than gas escaping from a high-pressure line, whose volume takes more time to mix with the surrounding air.

> **SAFETY TIP**
> An operator should not leave a gas line break until all safety precautions have been taken.

If a gas line is ruptured in a heavy traffic area, an operator's reaction may be to shut off the backhoe and get help. Doing so, however, is dangerous for two reasons. First, if the backhoe is equipped with a turbocharger, shutting off the backhoe without a cooldown period can cause the turbo to get extremely hot without the oil and fan to cool it. It can get hot enough to ignite the stagnant gas in the air. Second, if the backhoe is shut off immediately after the gas line is ruptured, the bucket will probably be in the way of pinching off the line or any other work necessary to stop the flow of gas. In order to move the bucket to do the necessary work, the backhoe must be re-started, and starting a backhoe produces a surge of static electricity that, under the right conditions, can ignite the surrounding gas.

> **SAFETY TIP**
> When gas escapes from a break in a line, the gas becomes flammable when a ratio of four parts gas to 100 parts air is reached.

Therefore, when a rupture in a gas line is suspected, whether by smell or by the sound of the gas escaping from the line, the operator should lift the bucket out of the trench and move the backhoe completely away from the area. The next step is to notify the proper personnel. A foreman or a superintendent should have the appropriate telephone numbers to call to have the line shut down and repaired.

If an operator is digging alone, which should never be the case, the gas company should be called immediately. If the operator doesn't have the number of the gas company, 911 should be dialed. The operator will direct the proper personnel to the location.

After the telephone calls have been made, public safety should be the operator's primary concern. If an operator is in a heavy traffic area, he or she should be careful not to compound the problem and perhaps cause a traffic accident by trying to close down lanes or divert traffic without the proper delineation. Pedestrian traffic should be kept as far away as possible.

The operator also needs to be careful not to spend lengthy periods of time near the escaping gas. If the operator starts to feel dizzy or nauseous, or if blurred vision is experienced, the operator should move to a safe distance, sit down, and wait for help.

6.6.3.1 Making Contact in a Residential Area

If contact is made with a gas line in a residential area, all the above mentioned precautions should be observed plus one more.

A backhoe exerts a great deal of force when digging. When contact is made with a gas line in front of a residence, the operator should check the damage very closely. If gas is escaping from the point of break, the procedure describe above should be followed. An operator also needs to watch for any extra slack in a gas line where the bucket made contact but there is no sign of a leak.

Steel gas lines don't stretch like plastic conduits. If there is slack in the gas line and it did not pull apart to cause the slack on one side, then chances are it came apart on the opposite side, causing the slack.

Chapter Six: Utilities

If this is the case, there is a good possibility that the force exerted by the backhoe has pulled the gas line apart inside the structure of the residence or under the foundation slab.

This type of gas leak is the most dangerous and can go undetected for a long period of time. Gas that is released under a house can hide in pockets in the ground or under the foundation slab, out of range of smell.

It is crucial, therefore, that when the operator makes contact with a gas line in front of a residence, work does not proceed as though nothing happened. The operator should check the residence that the gas line feeds. If necessary, the operator should knock on the door, explain the situation to the residents, and communicate concerns honestly. If anyone has any questions or concerns, the gas company should be called.

6.6.4 Making Contact with Telephone Lines

Breaking a telephone line or the conduit that it is carried in is, unfortunately, not as easily recognized as the rupture of a gas line. With telephone lines there are degrees of breakage. Once damage to the line or conduit is suspected, the operator should take the necessary steps to keep the damage to a minimum.

> *An operator should keep an eye on the spoils for any change in the material.*

If, when digging to locate a telephone line, an operator brings up a broken piece of conduit in the bucket, it can be assumed that there is a broken length of conduit in the trench. The operator should stop digging and investigate. The operator can use a shovel to clear bulk material. A nonmetallic object, like a piece of lath or a small stick, should be used to clean the dirt from inside the broken conduit to determine if the line inside was damaged. An operator should not use fingers to clean dirt away from inside a conduit until the operator is absolutely sure what the material is. There have been many instances when an electrical line has been mistaken for a telephone cable.

Once the broken conduit has been cleaned and identified as telephone, the operator should run a hand lightly over the cable to feel for nicks or severe kinks, being careful not to slice his or her fingers on the thin metal sheathing if there is a tear in the cable.

Operators should also keep in mind that most telephone conduits are pressurized with 10 psi of dry nitrogen gas to keep moisture out of the conduit and to alert the telephone company of a break in the line. When examining the cable, operators should be aware that the smallest hole will release the gas from the line. If there is any doubt as to the damage of the conduit, a little hand soap dissolved in a bucket of water and spread over the area will reveal any leaks by the appearance of bubbles.

When a conduit has been broken and the teeth have made contact with the cable inside, there are usually a couple of inches of slack in the cable before the actual cable starts to stretch and become damaged.

Making Contact with Utilities

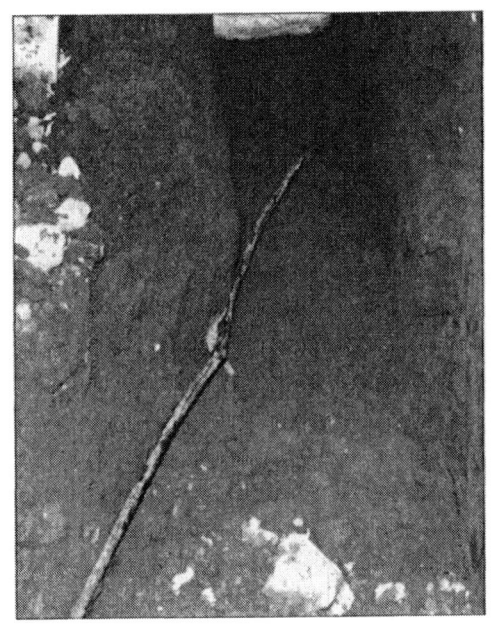

Direct burial telephone line is very difficult to detect.

With direct burial cable there is no conduit and therefore no slack.

The only advantage an operator can have is to dig the trench a couple of inches wider than the bucket in the vicinity of the location mark. This extra width will provide a little slack in the trench. An operator may not feel the cable with the bucket, but if the trench floor is watched closely, the operator will see the material around the cable start to move the instant the teeth make contact with the cable.

An operator should stop as soon as contact is made with the cable. Unless the operator can see the teeth and their position relative to the cable, pulling the

> **SAFETY TIP**
> Telephone lines carry 40 volts, raising to 120 volts when the line is ringing.

bucket out of the trench could easily tear or snap the cable. Whether an operator is working alone or with a laborer, it is well worth the time to determine the location of the teeth relative to the cable. Only then can an operator decide on the next bucket move.

Chapter Seven: Removals and Digouts

7.1 Introduction

Concrete and asphalt work are either digouts or removals. In a digout, an operator inspects the area to find the bad spots, digs them out, puts down a layer of base material into the patch, and lays a fresh lift of material on top of the entire area. When a larger area is so worn that it requires new material from the ground up, it is called a removal.

There are a number of guidelines that operators should follow when working with both asphalt and concrete in digouts and removals. First, the outriggers should be attached to the backhoe to stabilize it during operation. The outriggers are not backstops against which to pick up material. Second, the material should be kept in front of the backhoe. When the material starts to get too close, the operator should push the material back into the working area. Third, the operator should keep a clean level grade. If a piece is larger than the gap between the teeth, it should be moved.

Reaching into a pile of broken asphalt or concrete to retrieve a full bucket of material is relatively easy. The real skill is required at the beginning of the job when the operator has to figure out truck location, choose the best way to approach the removal or digout, avoid breaking pieces when pulling them from the edges and corners, and retrieving pieces that have fallen in difficult-to-retrieve places.

All problems need to be resolved quickly and efficiently.

7.1.1 Technique

The technique used for removals and digouts must be learned and practiced enough to be second nature so that the movements are one fluid motion. There are no special tricks that will instantly transform an operator into a removal/digout specialist. It takes an understanding of how broken pieces of asphalt and concrete react when bucket

Chapter Seven: Removals and Digouts

pressure is applied to them, and being able to manipulate the material so that taking a bucket full of material sets up the following pieces of material for the next pass.

> *The outriggers should be used to support the backhoe, not as backstops.*

Operators need to concentrate on three areas: The ability to pick up individual pieces of material without the aid of a backstop; sizing these individual pieces quickly to judge the best angle to approach them for removal; and leaving a clean grade with no chipped or broken edges.

7.2 Saving Edges

Whether doing a digout or a removal, edges are easy to break and almost impossible to repair. Broken edges occur when an operator doesn't pay attention to the placement of the loader bucket and outriggers, and when two pieces of solid material are forced against each other. When this occurs, the weakest point will break and this point will be the corner or the edge. An operator needs to make sure the weakest point is away from the edge, preferably at the center of the piece being removed.

7.2.1 Controlled Relief

The ability to save edges comes from understanding the reaction of concrete and asphalt under pressure. To achieve this an operator needs to know and practice controlled relief. The best way to explain controlled relief is through examples.

Controlled relief is the key to clean edges.

Saving Edges

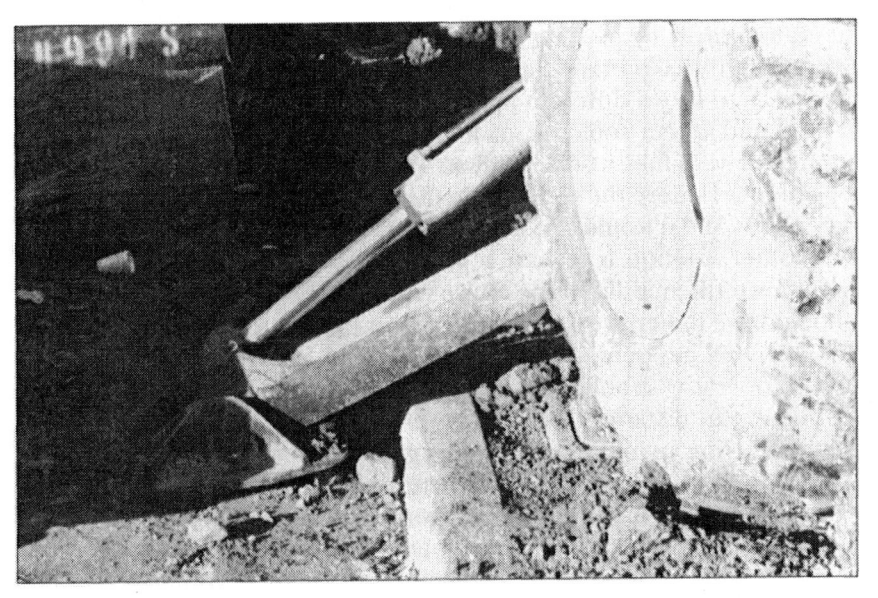

An operator should pay close attention to the location of the outriggers.

7.2.1.1 Example #1

The owners of a new grocery store discover they need to put in a new sewer line before the store can open. The new sewer will run straight down the middle of the parking lot, which is 6-inch-thick concrete.

The strip that needs to be removed in order to perform the work has been sawcut 15x150 feet. At the beginning of one end, a diagonal cut is made about 6 inches wide across one corner. This extra cut at the corner is called a pie and can be removed with the help of a screwdriver or shovel to give the teeth a starting point. The width of the removal allows the operator to set up the truck outside the removal line but still within reach.

The operator should center the backhoe in the middle of the removal, with the starting point about halfway out of the distance of the reach. The operator should then set the teeth into the starting point and curl the bucket in, locking the teeth against the concrete edge. (Rolling the bucket back to get more leverage would force the back of the bucket onto the edge behind it, probably causing it to break.) The operator then needs to boom straight up, using the bucket curl to keep the teeth locked onto the piece. If the piece will not lift with boom pressure alone or if any outside edges start to lift, the operator should stop immediately. The removal's perime-

> **SAFETY TIP**
> Chipping can occur when an operator pries the first pieces of concrete out of a section, sending small concrete pieces into the eyes of an unprotected laborer. Personnel need to stay away from the area or wear eye protection.

Chapter Seven: Removals and Digouts

ter is locked in by existing concrete and trying to work out the front edge center pieces immediately puts the concrete in a bind. The operator needs to take a different approach.

A second sawcut could be made around the entire perimeter of the removal, a few inches inside the first. This second sawcut would allow the operator to remove the pieces from the center without disturbing the outside edges, and it would also absorb the abuse, acting like a cushion.

Another solution is to have a jackhammer or stomper break a relief strip down the middle of the entire removal. If this occurs, the operator should have the entire front edge broken for a bigger starting point. As long as relief has been created in the removal, work can proceed.

The operator should then set back up in the center of the removal at the proper distance from the front edge, bringing the teeth back into the starting point, locking them against the edge, and booming up. If the method of release was the stomper or jackhammer, the piece should lift right up. If not, the operator should pull the teeth back out, curl the bucket, and give the edge where the starting point is a solid rap with the back of the bucket.

Operators need to use caution when using this technique because concrete that is still in its original poured position on the ground is completely without cushion when it is hit with the bucket. It's all right to use the back of the bucket to break pieces that have been pulled up and propped on one end, but consistently hitting concrete that is lying flat on the ground will cause shock damage to everything, includ-

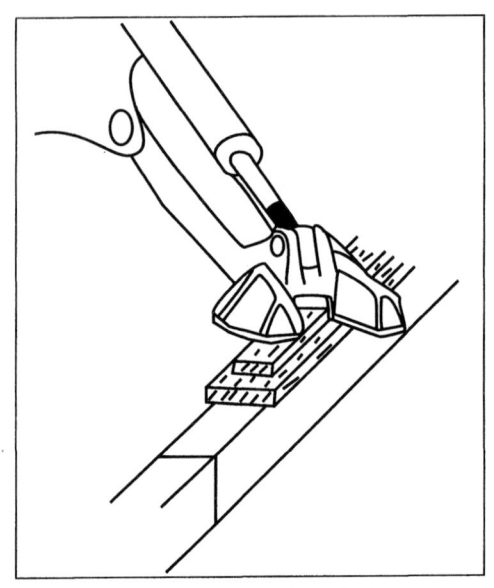

An operator can use wood as a cushion when placing an outrigger on a curb.

ing the back of the bucket and the swing bushings.

The operator should then pull the piece just hit out of the grade and place it in the truck. The entire front edge needs to be removed before the center is taken out. If there is a relief strip down the middle, the next bite should be from the center of the removal, otherwise work will be done from the front edge back, with the operator working from the center out.

Once the front edge is gone, the operator needs to establish a pattern for the rest of the removal. This pattern should consist of removing the center, or as close to it as possible, pulling the pieces away from the edge, and performing the final cleanup.

When the operator positions the bucket to remove a piece of concrete that is against the edge, before pulling directly up the operator should pull flat along the grade toward the machine until it is free of the edge. After being freed from the edge, the piece can be lifted by the bucket.

7.2.1.2 Example #2

The second example involves a sidewalk removal that is 4 feet wide and poured against curb-and-gutter. If the size of the area allows it, the operator should set up straddling the sidewalk. If space is a problem, the operator will need to use good judgment, always keeping the loader and outrigger placement in mind. For this example, there is plenty of room and so the setup is straight on.

The sidewalk is adding support to the back of the curb, so the operator knows that the loader bucket, placed flat, and the outriggers, when placed evenly on top of the curb, will be fine.

When an operator has to place an outrigger in the gutter, careful attention must be paid to the wrist of the outrigger. When removals are performed with the backhoe in a low profile and close to the ground, the wrist can damage the nose of the curb.

Setting the backhoe up at a higher profile increases the distance from the wrist to the nose of the curb, eliminating the problem.

To remove the sidewalk from a straddling position, an operator should take the following steps:

1. The teeth should be brought to the edge or starting point on the sidewalk and locked on, using only the bucket curl. Before an operator does any lifting, the swing should be used to pull and slide the sidewalk away from the curb. Immediately pulling the sidewalk straight up will break the back of the curb. If the sidewalk will not pull away, the operator should give the sidewalk a quick rap with the back of the bucket about 2 feet in from the edge and try again.

2. After the sidewalk is away from the curb, the operator should check for a keyway. If the sidewalk contains a keyway, the operator will need to pull the sidewalk away from the curb until the entire keyway is free along the piece of sidewalk before picking it up to place it in the truck.

Chapter Seven: Removals and Digouts

3. The operator should break large pieces into the proper size after lifting one end up. Lifting one end up only to set it back down again may seem like an extra step to operators, but the concrete should break much easier and when it does break, the pieces should be larger and easier to remove.

4. An operator's last pass will be the cleanup pass, retrieving any small or leftover pieces.

7.3   Picking Up from a Flat Surface

All operators need to learn how to pick up a piece of concrete or asphalt off a flat surface. Just as with bulk removals, an operator must be able to get the teeth under the object to pick it up. Without a backstop to support the piece while sliding the bucket under it, an operator must rely on the teeth and movements of the bucket to manipulate the piece of material off the flat surface into the bucket.

The operator should set up the backhoe so that the material is about three-fourths the distance of the overall reach. The operator should also keep an eye on the perimeter edges. If there is an edge or corner that has a lip or is canted so that a tooth can get under it, that piece should be positioned with that edge away from the operator.

With the teeth flat on the surface the operator is retrieving the piece from, the bucket should be drawn back very slowly under the piece until it starts to slide. Once the piece starts to slide, the area of less resistance has become the flat surface and it will slide along the top indefinitely.

> *If an operator uses the backhoe bucket to snap a piece, the maneuver should be done with the full back of the bucket, not just the edge.*

To see if the teeth are hooked under the piece, the operator should curl the bucket very slowly. The bucket only needs to be curled an inch or so to be successful; the operator will know if it has been a success or failure by whether or not the piece raises when the bucket is curled. If the technique was correct and the operator was still unable to get the teeth under the piece, the operator should not waste time doing the same thing on the same side because the piece will slide every time.

The operator should either spin the piece to another side or, if the piece slid considerably, push it back to its original starting point. Starting off the same way with the teeth flat, the operator should draw the bucket in and try to get under the piece, slowly incorporating the swing into the action. The goal is to get one of the corners to catch while the piece is slowly rotating. It makes no difference which side the operator swings to, but it is important that it is done very slowly or the operator may spin the piece right off the teeth.

Whether using the swing or straight-on approach, once the operator succeeds in getting a tooth under the piece, the bucket cylinder should be left alone and the piece picked up with the boom and crowd. The operator should boom up very slowly until it looks like the tooth may lose its place under the piece because of the change in angle.

At that point, the operator should slowly bring the crowd arm in and regain the tooth's position on the piece. If the piece starts to slide, the operator should stop because the operator needs to change the angle of either the boom or the bucket to place more weight on the edge of the piece that is on the ground. In this case the boom needs to come straight up until either the piece is standing vertical or the teeth look like they may slide off. If the operator thinks the teeth look like they are getting ready to loosen their grip, then the operator should slowly open the bucket while drawing in on the crowd arm to compensate. It is important that these two moves are done simultaneously, slowly, and smoothly. If executed properly, the piece should not move at all.

> *Before an operator lifts a piece, the swing should be used to pull the sidewalk away from the back of the curb.*

At this point, moving the piece even the slightest bit risks it sliding back down and possibly falling flat. For the operator to get the piece from a standing position, these steps should be followed. First, with the piece in a standing position, the operator needs to manipulate the bucket floor to a parallel position, with the piece as high to the bucket collar as possible. Next, the operator should boom straight down until the piece jams against the craw of the bucket. If the piece isn't long enough to do so, the operator should boom down until the teeth make contact with the ground. The bucket should not be curled at this point.

Watching only the point where the piece makes contact with the ground, the operator should slowly bring the teeth to the ground and gently start to crowd out. The instant the operator incorporates the crowd, the bucket should be curled. The operator must also keep the bucket angle matching that of the piece as it is lowered.

When the operator has boomed down to a point where the back of the bucket is touching the ground, the bucket should be slowly curled. If more than 50 percent of the piece is in the bucket, it should stay as the operator continues to curl. If the piece starts to slide out, the operator should bring it to a vertical position again and, keeping the piece on its edge, roll the bucket to where the tips of the teeth are supporting the piece and the bucket sides are level with the ground. Again, the operator should slowly boom straight down, watching the piece. When the bucket is about a foot from the ground, the operator should crowd out a little and quickly boom straight down, allowing the piece to fall toward the bucket. The instant the piece starts to fall and enters the bucket, the operator should quickly curl the bucket to a holding position.

Chapter Seven: Removals and Digouts

7.4 Picking Up a Piece from the Grade

Picking up a piece of material out of the working grade is easier than picking a piece off a flat surface. Because the subgrade acts as a backstop, choosing a specific corner to pick up is not necessary. The operator's primary concern should be making sure the piece is a suitable size and can be picked up lengthwise.

The size of the piece to be picked up depends on where it is going after it is loaded into the truck. If the material is going to a plant to be crushed, it is preferable that the pieces be no larger than 2 or 3 feet square. If the pieces are going to a stockpile, then the size of the pieces depends on the driver of the truck. Keep in mind that the smaller the pieces, the more the operator can fit into the truck.

Picking up a piece off the working grade should be one smooth movement. The bucket should approach the piece opened slightly more than in the flat position. The operator should bring the teeth under the edge of the piece and immediately boom up, simultaneously crowding in a little so the teeth don't slide off the piece. As the operator booms and crowds in to compensate, the bucket should be opened so that the bucket's sides match the angle of the bottom of the piece. When the piece hits the craw of the bucket or when the teeth make contact with the ground, the operator should boom down to force the end of the piece into the ground, which, in turn, will force the piece deeper into the craw of the bucket.

> *The smaller the pieces are, the more an operator can fit into the truck.*

The operator then needs to curl the bucket by rolling it back on the top of the subgrade while keeping it in one place. This can be done by crowding back a little to keep from forcing the piece into the grade and by booming down while curling the bucket.

If the operator is able to wedge the piece into the craw of the bucket, the bucket should be curled until the operator feels the bucket grip the piece. If the piece isn't big enough to reach the craw, then the operator will have to continue rolling the bucket back until the piece comes to rest on the teeth.

7.5 Differences in Flat Surfaces vs. the Grade

There are a few differences between picking material up out of a working grade and picking it up off a flat surface. Because the subgrade under the piece of asphalt or concrete will be road base or dirt, both of which are suitable for use as a backstop and will allow a corner to dig in, the subgrade will hold the down-side edge with sufficient force to slide the teeth under. This allows the operator to skip unnecessary moves.

Differences in Flat Surfaces vs. the Grade

Picking up concrete takes practice and patience.

Chapter Seven: Removals and Digouts

Another important difference is with backhoe movement. Because the operator is moving slowly and pulling levers individually, the movements may appear to be jerky. However, the operation should be as smooth as trench digging.

7.6 Picking Up Uneven Pieces

In theory, the perfect piece of material to pick up would be the same distance from the craw of the bucket to the tips of the teeth and just shy of the inside width of the sides of the bucket. But just because every piece of material does not have one flat edge that can be used for standing it up doesn't mean that the operator needs to break every piece to create a flat side. It does mean, however, that the operator needs to acquire the ability to pick up pieces of any shape or size without having to rely on a flat side.

A piece of concrete or asphalt that has been stood up to balance on its flat edge will fall either forward or backward when released. If that flat edge is removed, it will fall forward, backward, or to the side. It is important for an operator to be able to anticipate this direction. An example illustrates why.

The operator takes a piece of concrete as big as the bucket in the shape of a triangle and stands it on its tip. The operator positions the backhoe so that the piece is centered in front of the backhoe. When the operator places the bucket flat on top of the piece and booms straight down, the piece is forced to fall to one side or the other by the pressure from the boom. Some might think that because the piece is in the center of the backhoe and equal force is being applied to all three sides, the piece would remain in place and the backhoe would lift because of the boom pressure. If the backhoe was a solid piece of steel from the ground to the bucket and it rested on the top of the piece of concrete, this might occur. But slack is created by the hydraulics, pins, and bushings, and when pressure is applied to the top of the triangle, the accumulated slack allows the boom to shift in weight from the center of the piece to either side.

> *Proper maintenance is the key to a tight backhoe loader.*

This example applies to field removals each time an uneven piece of concrete or asphalt is picked up. When an ordinary piece of concrete or asphalt is picked up and pressure is applied to the piece, the piece will roll to whatever side offers the least resistance.

The piece doesn't have to be in a vertical position to roll when the weight is applied to it. When the operator first gets the teeth under a piece that is lying flat, the piece will react the same way: It will roll to whichever side offers the least resistance. When an operator is about to pick up an uneven piece of material, the operator will need to anticipate the direction the piece is likely to roll and begin compensating with the swing cylinder so that the piece will stay straight while it stands and is eventually picked up.

An operator can do all this by closely watching the piece as the teeth get under it. When it starts to roll to one side or the other, the operator should immediately start to compensate to keep the piece straight.

7.7 Digouts

A digout entails the removal of potholes. An operator should set up on a digout like setting up for anything else, with the beginning of the digout about three-quarters out of the overall reach and with the backhoe centered. The truck being loaded should be about 5 or 6 feet outside the outrigger in a parallel position, with the head of the truck facing the same direction as the machine. An operator should never have the truck turned so that the material is brought over the cab of the truck. Whenever possible, the truck should be loaded from front to back.

> *When possible, the operator should load the truck from front to back.*

The perimeter of the digout will either be sawcut or jackhammered. In either case, the operator needs a starting point to get into the material to lift it out. If the perimeter is sawcut, it will generally have a double cut on the front edge. A double cut is a set of cuts a few inches apart that can be popped out with a shovel or digging bar, leaving an access point for the teeth. If the perimeter is jackhammered, the front edge will be broken up again allowing access for the teeth.

If the perimeter has been cut but no starting point has been left for

Digouts.

Chapter Seven: Removals and Digouts

the teeth, there are a couple of things an operator can do. If it is hot on the day of the digout, the operator should bring the teeth to the very edge of the digout, place them just short of the front perimeter line, and boom straight down until the backhoe is off the ground and all of the weight is on the teeth. The operator should then lift the loader about a foot off the ground and slowly swing the backhoe from left to right. This left-to-right motion should be continued until the teeth penetrate and work through the asphalt.

> *An operator should not ruin clean edges by gouging them with the teeth.*

On cold mornings, the teeth will not penetrate the surface. In this case, the operator should position the backhoe with the front digout line directly in front. Lifting the loader about 2 feet into the air and with the outside tooth in line with the digout perimeter line, the operator should lower the opposite side outrigger until it can no longer raise the backhoe.

At this point, the other outrigger should be lowered until it just makes contact with the ground for stability. The loader bucket should also be lowered to ground level with the operator making sure that the corner of the loader that meets the ground is inside the digout perimeter. As the operator starts to move the bucket, the backhoe will move back and forth slightly, gouging the asphalt with the loader.

With the outside tooth at the beginning of the front line, the operator should make the first pass a ghosting pass, with the tooth riding just above the line to make sure it is straight as the bucket is pulled in. The first contact pass must be perfectly straight, with the tooth scribing a line just inside the front perimeter line. Once a scribe pass has been made, it is very difficult to make corrections.

Starting at the farthest corner and pulling a straight line toward the backhoe, the operator should scribe about a half inch at a time until either the subgrade below has been reached, or until the operator thinks the teeth can be worked into the remaining asphalt from the digout loading position. Although this will take time and it will wear out the outside tooth, it is easier than using the backhoe as a jackhammer.

Once the starting point has been opened, the operator should remove the material while keeping the removal rules in mind: a clean grade, clean edges without any chips, and piling material where it can be easily accessed.

7.7.1 Finishing the Digout

As the operator nears the last edge, the backhoe's moves need to be geared toward a clean finish. The final setup should place the last edge of the digout about halfway toward the backhoe in a position where the pieces that flop over the edge can be pushed back into the grade for retrieval.

When the last of the large pieces of material has been removed and there are just small chunks remaining, the operator should bring the bucket to a flat position with the bucket floor parallel to the grade. Starting back about a foot past the small chunks and with the bucket flat on the ground, the operator should slide the bucket along the top of the grade, picking up the chunks until contact is made with the last edge; the chunks in the bucket should be dumped.

The operator should return to the grade with the bucket opened to the full position, and setting the teeth against the last edge, press down into the grade. The operator should dig about an inch below the grade level to break any pieces away from the edge. The operator should then raise the teeth back to grade, push the material that is behind the teeth back about a foot, and raise the bucket over and behind the material that is ready to be picked up.

With the bucket about halfway between the open and flat positions, the operator should bring the material within 2 or 3 inches of the edge, boom straight down into the grade about 2 inches, and curl the bucket an inch or two. Bringing the crowd arm in towards the backhoe and forcing the material on the teeth back a little farther to a holding position on the cutting edge, the operator should continue to pull in until contact is made with the edge of the asphalt. Crowding back out until the teeth clear the edge and dumping the bucket should leave a clean edge. If it doesn't, the operator should try again until it does.

7.8 Removals

Generally, a removal entails the removal of a long stretch of asphalt, such as an entire side of a street. Removals may include removing a section wider than what can be reached by the backhoe in one pass.

Before work can begin, an operator needs to answer an important question. If the truck is positioned outside the removal area, is it possible to reach the entire section with the backhoe, leave a clean grade, and still load the truck? If the answer is yes, that's what should be done. If the answer is no, the operator needs to figure out the quickest and most efficient way to get the material from the ground and into the truck. For example, an area to be removed is 75 feet long and 25 feet wide. With the dump truck set up about a foot from the outside perimeter line, the operator will need to place the outrigger no farther than 6 or 7 feet from the truck to assure a proper reach. Because of the extreme width of the removal area, swinging the boom to the far side of the removal will put the teeth a few feet short of the far outside perimeter line.

All ground personnel should be kept away from the swing radius.

With the loader bucket planted firmly on the ground, the operator could pivot over to the far side of the removal, take a bucket full of material, pivot back close enough to the truck to dump it, and then pivot back for another bucket full. This is a good method to use if the removal area is only about 25 feet long. But with the area being longer, moving the backhoe twice for every bucket full is a waste of time.

By reaching over to the far side of the removal from the first setup point by the truck, the operator determined that the bucket was a few feet short of the far perimeter line. Since the removal is so long, it is worth the operator's time to set up on the far side of the removal, setting up the truck in a position a couple of feet more than what is necessary to reach from the opposite side and remove that entire section of asphalt.

> *The backhoe bucket must make full tooth contact on a piece to avoid bending the bucket's front cutting edge.*

The operator should not be trying to split the section in half. Instead, enough material should be taken to accomplish two things. First, when the operator sets up on the opposite side to do the bulk of the removals, the backhoe should be able to reach all of the remaining material. Second, the operator should be able to judge the amount of material being taken on the far side to fill the truck.

It defeats the purpose to remove the strip of asphalt on the far side and have only half a truck of material, only to find that the backhoe must be folded up and repositioned at the beginning of the other side in order to finish loading the same truck. The advantage of this method is that any repositioning of the backhoe can be done after the truck is loaded or when a new truck is pulling into position.

> *On a warm day, a loader bucket's cutting edge and pins will dig into asphalt.*

7.9 Picking Up Individual Pieces with the Loader Bucket

Picking up individual pieces of concrete or asphalt in the loader bucket is a little like flying blind. An operator can't see through the loader bucket to find out what is going on when trying to pick up a piece; the operator has to wait until the loader bucket is curled to see whether or not the end of the piece rolls up with the bucket, indicating a successful attempt.

There are a few techniques that can help an operator when picking up individual pieces. For example, to pick up a 5x5-foot piece of concrete off a flat, smooth surface such as asphalt, it is necessary for the operator to get the cutting edge under the piece.

There are two ways to accomplish this. The first is to move the backhoe loader back about 30 feet from the piece and put the machine in second gear. At about 10 m.p.h, the operator should hit the piece with the loader riding on the ground in the flat position. Once the piece

Picking Up Individual Pieces with the Loader Bucket

When figuring truck location, the operator should never plan on loading over the cab of a truck.

has been hit, it will go into the loader. If the operator comes to an abrupt stop after the piece is hit, the momentum gained upon impact will allow the piece to slide out of the bucket.

Keeping the piece in the loader bucket long enough to pick it up is the key. The trick lies in stopping the backhoe with the piece still in the loader. The operator needs to position the backhoe loader so that when the piece is impacted, it is as far to one side of the loader as possible. When the piece is first impacted, the operator will feel it go into the loader bucket. When this happens, the operator should lay heavily on the turning brake, opposite to the side where the piece is in the loader. The operator should not lay so heavily on the brake so that the machine is on two wheels, but heavily enough to use the turning of the backhoe loader to keep the piece wedged in the corner of the loader bucket while coming to a complete stop.

The second way is to position the backhoe loader at a dead stop with the piece off to one side of the loader, resting against the cutting edge. The operator should apply enough down pressure to the loader to get the front wheels off the ground. The operator should then push on the turning brake hard enough to hold the tire of whichever side the piece is against on the loader edge. Still holding the turning brake, the operator should accelerate hard. The backhoe will pivot on the tire that is stopped, allowing the cutting edge to slice under the piece. The operator should stop accelerating when the piece is resting against the opposite end of the loader from where it started.

The next step is to make sure the edge of the loader bucket is under the piece. The operator should bring the backhoe loader to a complete stop, slowly raise the loader bucket, and lean over far enough to look under the cutting edge.

Chapter Seven: Removals and Digouts

As the operator is lifting the piece of concrete to see if the cutting edge is under it, the operator should also be watching where the concrete is making contact with the ground. With the piece on the edge of the loader, the piece will slide from its contact point when the loader is lowered. When the loader is raised, the piece will stay on its contact point.

With the backhoe loader in low gear and with the piece on the lip of the cutting edge, the operator should raise the loader bucket and be prepared to move the machine forward to keep the piece on its contact point if it starts to slide. When the operator reaches the point where the piece is standing as vertical as possible without flipping over, the operator should start to roll the bucket over, still watching the contact point, until the loader bucket floor is parallel with the piece. The loader should be lowered in its curled-over position until the piece makes contact with the back of the bucket.

The operator should then place the backhoe loader in reverse and, without moving, lower and curl the loader bucket simultaneously to keep the piece parallel with the bucket floor. If the piece starts to slide, lowering the loader a little faster should stop it. If not, the loader should be raised until the sliding stops. The operator may have

An operator needs to remember that practice and patience are vital when operating a backhoe loader.

to place the backhoe loader in a forward gear and move up a little if the sliding is drastic.

While the loader bucket is curled and lowered to the ground, the piece must stay parallel to the bucket floor and it also must be kept against the rear wall of the loader bucket. The operator should continue curling and lowering the piece until the loader bucket floor is lying flat on the ground. The operator should now be able to curl the loader bucket slowly, while making sure the piece remains against the rear wall.

With the loader bucket flat on the ground, being curled, the piece may act as if it is going to fall back out. As the operator continues to curl the loader bucket, the edge of the piece that is in the loader will wedge itself against the rear wall as it starts to teeter out. As the operator continues to curl and reach the over center point of the piece in the loader, the piece will slide back to the rear of the loader bucket in a comfortable holding position.

7.10 Picking Up Individual Pieces Out of the Grade

The important difference between picking up a piece off a flat surface and picking it up out of a working grade is that the grade material will provide a backstop to pick up against. This pickup technique is a simplified version of the method above.

Using the same example, except replacing the asphalt with dirt, the operator should approach the piece with the loader bucket in the 4 o'clock position. After pushing the loader bucket edge through the grade and picking up the edge of the piece with the bucket lip, the operator should lift the edge of the piece until it is at the same angle as the loader.

With the backhoe loader in low gear, the operator should drive slowly forward, forcing the piece deeper into the loader. As the operator is driving forward, the operator is going to need to lower the loader bucket to compensate for the change in angle of the piece. It is important that the piece stay at the same angle as the loader bucket at all times. When the piece is as far into the bucket as it will go, the operator should lower the loader bucket to the ground and slowly curl it up.

At this point, there must be no forward pressure on the piece or it will flop out. As a safeguard, the operator may want to back up slightly before curling the loader to hold the piece.

Chapter Eight:
Jumping a Trench

8.1   Front-first Jumping on the Go

Front-first jumping can save an operator time. The option can be exercised if the conditions are right. These conditions are:

1. There must be at least an 18-inch bucket on the backhoe. This is necessary because the operator will not have the loader edge to stabilize the backhoe when it becomes airborne as the trench is crossed.

2. An operator should not do this procedure on a trench wider than the distance from the center of the front tires to the bottom of the loader pins. The loader will be needed to support the weight of the front end of the backhoe loader as the trench is crossed.

3. The on-the-go method is designed for four-lever controls with the operator using a hand for the controls. Although this method can be done with wobble sticks, the operator will need to use an entire arm to maneuver both controls simultaneously.

Because of the complexity and time involved with jumping, an operator should do the maneuver only when all other options have been exhausted.

The following example assumes these conditions. Electrical conduit has been placed in an 18-inch-wide trench, and the specifications call for a layer of sand over the conduit before backfilling. But the dump truck driver got confused and dumped the load of sand on the wrong side of the trench. The operator decides to go over the trench, instead of around it, to save time. The object is to cross the trench in one smooth motion, **without stopping the backhoe.** Hence the term "on the go."

After taking a loader bucket full of sand and approaching the trench, the operator needs to position the boom-and-crowd in preparation for the jump. With the boom-and-crowd in their positions, and with the bucket riding about a foot off the ground, the operator should bring the full loader bucket to the far side of the trench and

Chapter Eight: Jumping a Trench

set it down with enough pressure to lift the front tires off the ground.

The operator should drive forward and, while riding on the front loader pins, allow the front tires to slide over the trench. The operator should continue to drive forward until the rear tires approach the side of the trench, then lift the loader bucket about 8 inches off the ground and allow the front tires to take the loader weight.

> *The front tires must be fully inflated because the weight transfers to the front end of the machine when the rear is elevated.*

With one hand on the steering wheel and the other working the boom-and-crowd levers, the operator should drop the boom, lift and push the backhoe over the trench. If the backhoe starts to sway halfway through the push, the operator should take his or her hand off the steering wheel and, with the other hand still working the backhoe levers, lower the loader just enough to let the loader corners stabilize the backhoe. The operator should not let the loader pins come in contact with the ground. If that happens, the backhoe will come to a complete stop.

When the rear tires reach the other side of the trench, the operator should fold up the boom-and-crowd and continue.

8.2 Jumping a Trench at 45 Degrees

To jump a trench at 45 degrees, the operator should approach the trench with the loader bucket a few inches off the ground, and drive the loader bucket onto the trench at a 45-degree angle until the trench splits the loader in half. The full weight of the backhoe loader should be placed on the front of the loader until the front wheels are off the ground.

> *The operator should make sure the placement of the backhoe bucket is far enough away from the trench wall so that the bucket won't collapse the wall when the backhoe loader is elevated.*

With both outriggers set just above ground level, the backhoe bucket should then be set no closer than 2 feet to the edge of the trench, on the same side as the rear tires. The operator should apply down pressure to the boom and then lift the backhoe about 6 inches off the ground, allowing the loader to take the full weight of the machine.

The operator should swing sideways and crowd out slightly until the backhoe has come to a straddling position over the trench and both outriggers can be set on either side of the trench safely, without danger of collapsing a trench wall. The weight of the backhoe loader should be set on the outriggers and the boom moved to the other side of the trench.

The boom should be set down no closer than 2 feet to the side of the trench on the opposite side. The operator should then boom down until the rear tires are clear of the ground, then swing over and crowd in slightly until the backhoe comes to rest on the other side of the trench.

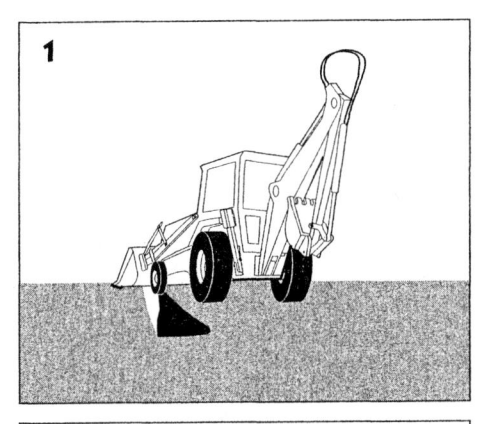

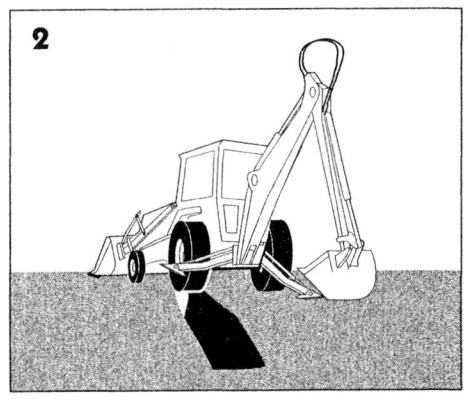

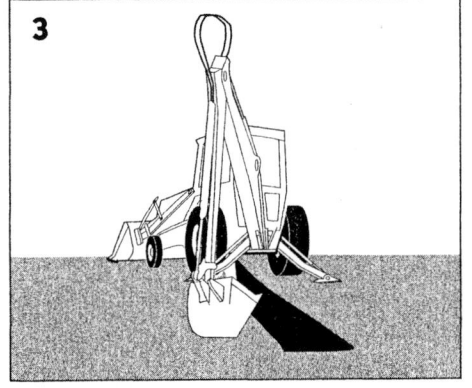

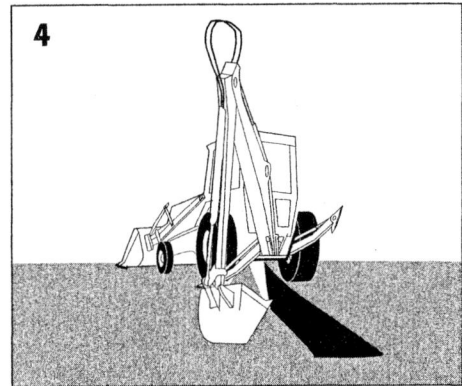

Jumping a trench at a 45-degree angle.

The operator should not set down until he or she is sure the outrigger closest to the trench is out of danger of collapsing the trench wall if weight is placed upon it. The operator can then fold up the outriggers and engage the reverser, backing off the trench with the loader still elevating the front tires until they are clear of the trench.

8.3 Jumping a Trench Straight on

Pushing or pulling the backhoe loader over any obstacle straight on is always much easier with a wide backhoe bucket. If an operator is jumping a trench using a 1-foot bucket, the loader bucket should be kept low enough to the ground so that when the operator lifts the backhoe in preparation to push, the loader will keep the backhoe from swaying from one side to the other, making the push difficult. This position is especially helpful when the backhoe must be kept stabilized in order to jump over a fixed obstacle like concrete form boards.

Chapter Eight: Jumping a Trench

The operator should approach the trench at a 90-degree angle and place the loader on the far side of the trench, with the front tires still on the side. If the trench is so wide that the operator can't lay the loader bucket flat on one side with the front tires on the other side, then, as the operator approaches the trench with the loader bucket in the flat position, the cutting edge or lip of the loader should be placed on the far side of the trench and enough down pressure applied to the loader to lift the front wheels off the ground. The operator should then slowly drive forward, using the loader to carry the front wheels over the trench.

> **SAFETY TIP**
> The safety belt will keep the operator in a secure position if the backhoe loader starts to sway.

If, after the operator has pushed the loader over the trench, the front tires fall short of making contact with the far edge of the trench, then the operator has two options. If the material on the far side of the trench is hard, the operator can roll the loader bucket slightly to keep the cutting edge from digging in, but not enough to allow the loader pins to drag. The operator should apply

Jumping a trench straight on.

Jumping a Trench Straight on

enough down pressure to the loader bucket so that as the backhoe loader is pushed the rest of the way over the trench, the front loader bucket slides along the top of the dirt, allowing completion of the push.

The other method can only be used if the tires are falling about a foot shy of the opposite trench side. The operator should turn around and place the crowd-and-boom in the position to push and sink the bucket teeth into the ground. Lifting the backhoe loader about a foot and applying a little crowd-out pressure, the operator should start very slowly and curl the backhoe bucket under as the crowd-out pressure is gradually increased. If the operator curls the loader bucket before applying a little crowd pressure with the bucket teeth in the grade, the operator might draw the loader across, putting the operator and the backhoe loader in a difficult situation: How would the operator return the loader to the flat position with the tires suspended in midair over the trench?

The operator should continue crowding out and curling the bucket under until the loader bottoms out and the front tires are safely on the other side. With the backhoe still in the air, the operator should lift the loader bucket to transfer the weight from the loader to the front tires. Pushing should continue until the backhoe loader is over the trench.

The operator should position the rear tires as close to the side of the trench as possible. However, the tires should not be so close that the backhoe is at risk of falling into the trench if there is a cave-in.

With the rear wheels positioned on the edge of the trench and with the front wheels solidly resting on the other side of the trench, the operator should raise the loader bucket to within about a foot of the ground, turn around, and prepare to push the backhoe over the trench. Crowd arm positioning is crucial at this point. An operator who booms down in an attempt to lift the backhoe with the crowd arm positioned too far in creates an overcenter weight problem, putting the crowd arm in a stall and making movement impossible. This position can also cause problems when the operator runs out of crowd arm halfway over the trench and has to pull the arm back, only to reposition it to try again.

If an operator is not sure of the position the crowd arm needs to be in before the push, the trial-and-error method should be used. The operator should start in close, lifting up the backhoe by booming down. If the backhoe will not move when the crowd arm is initiated, the operator should boom up again, move the crowd arm out a little, boom down, and try again. This trial-and-error method should be used until the backhoe moves forward.

When the operator is confident the trench can be crossed in one smooth motion, the loader bucket should be kept a few inches off the ground. Using the crowd for the actual pushing power and the boom for the height adjustment, the operator should push until the backhoe is safely to the other side of the trench.

Chapter Eight: Jumping a Trench

8.4 Jumping Back First

To jump a trench back first, the operator should back up to the trench until the rear tires are at a comfortable distance to the trench and set out the outriggers for stability. Reaching out over the trench with the backhoe bucket and setting the teeth into the grade as far back as possible without bottoming out the crowd cylinder, the operator should apply a little down pressure to the boom and raise the outriggers about 6 inches.

The loader bucket should be set about 6 inches off the grade and the backhoe loader lifted up. After taking a quick look at the loader bucket to make sure it is not rubbing on the dirt, the operator should start pulling the machine over the trench. The operator's eyes should be on the front tires to make sure the front end is not pulled into the trench.

After the operator has pulled as far as the front tires will allow, the backhoe should be folded up, the outriggers pulled up, and enough down pressure applied to the front loader bucket to lift the front tires

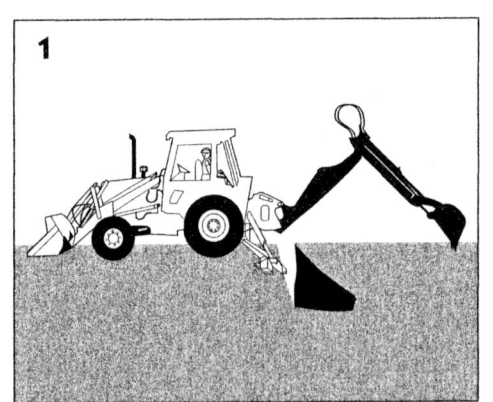

Jumping a trench back first.

off the ground. The loader bucket should be slightly rolled so the bulk of the weight is on the front cutting edge. The operator should then drive slowly backward until the front tires are over the grade on the side of the trench.

If the operator has jumped a trench back first that is too wide and the front tires won't reach the edge of the trench, the operator should stop and plant the outriggers. The operator should then turn around and reach out as far as possible and take a bucket full of material, holding it out as far as possible and suspended about a foot off the ground. This transfers as much weight as possible to the rear of the backhoe to offset the weight in the front. The operator should slowly raise the outriggers until they are just a few inches off the ground, and then very slowly lift the loader to make sure all of the weight has been transferred to the rear of the backhoe. With the front tires suspended over the trench, the operator should drive slowly backward until the tires are safely over the side of the trench; the outriggers should be set and the material released from the backhoe bucket.

> *It takes a great deal of force to pull a backhoe loader over a trench. If the operator sets the teeth in the grade too close to the trench wall, the wall will collapse as the operator starts to pull.*

If the operator has made the mistake of pulling the front end of the backhoe loader into a trench that was too wide, the above method can be used with a slight variation. With the outriggers down for stability, and with as much material as the bucket will hold, the operator should extend out all the way, including the extension if the backhoe has one. If the loader starts to raise from the trench, the operator is on the right track. If it doesn't, the operator needs to look back to be sure the backhoe bucket is no more than a foot off the ground. If it is, the operator should slowly start to raise the loader bucket into the air, transferring even more weight to the rear of the machine.

> **SAFETY TIP**
> Raising a loader bucket full of material will roll the machine on its side.

The operator shouldn't make the mistake of holding material in the loader as it is raised, thinking that the extra weight will help because it won't. It will, however, roll the backhoe on its side.

At some point the front tires will start to lift from the trench as the loader is being raised. The moment the tires start to move, the operator will know there is very little weight on the front end, and so the loader can stay where it is.

While reaching back with one hand to raise the backhoe bucket as needed to keep it off the ground, the operator should back away from the trench. The backhoe bucket is the only means of stability when you are backing up with the loader in the air. An operator should make any elevating adjustment to the backhoe bucket slowly and smoothly.

Chapter Eight: Jumping a Trench

8.5   Jumping Over an Obstacle

An operator will encounter any number of obstacles, including a concrete form board.

In this example, the operator should approach the form with the backhoe in a parallel position, about 2 feet away from the form. The loader bucket should be set at about a 5 o'clock position so that the weight of the front end is resting on the lip of the loader. The operator should apply pressure to the loader bucket until the front loader cylinders bottom out and the front tires are as high as they will go.

Placing one outrigger down for stability, the operator should move the boom to about half the distance of the overall reach of the backhoe and about 3 feet from the form. The operator should then boom straight down until the rear tires are higher than the form; the operator can then start to swing over.

> *If an operator is not comfortable jumping over an obstacle, an alternative should be found.*

The operator should keep an eye on the corner of the loader bucket that is closest to the form. When the backhoe begins to swing around and the loader starts to pivot, the operator will want to make sure that the corner of the loader bucket does not pivot right into the form.

When the backhoe loader is in mid-air over the form, the operator should set down the outrigger. It will eventually come to rest on the opposite side of the form about a foot from the ground; the operator will use this outrigger to slide on. The operator should continue swinging the backhoe over until the weight of the backhoe is on the outrigger that was placed near the ground.

At this point the weight of the backhoe loader will be overcenter, and the only thing keeping the machine from crushing the form is the outrigger that was put down on the far side. Because of the weight transfer, the operator will need to rely on the crowd to push the backhoe the rest of the way, sliding the outrigger across the dirt until either the opposite outrigger has cleared the form or the operator can continue to push a full 90 degrees to the form.

If the operator runs out of crowd and it can't be pushed a full 90 degrees to the form, the operator should place the other outrigger down and move the boom back out in front to push the rest of the way.

When the operator comes to rest in a position 90 degrees to the form board, the backhoe should be folded up and the outriggers slowly lifted. As this occurs, the operator will need to watch the front of the backhoe to make sure that nothing under the backhoe comes in contact with the form as the rear wheels approach the ground. When the rear wheels are resting on the ground and the outriggers are in their full up position, the operator should place the reverser in reverse and slowly move the backhoe away from the form board. The loader bucket should still be at its original 5 o'clock position. As the operator starts to back away from the form, the loader bucket should slide along the top of the material until the front wheels are clear of the form and the operator can set the front wheels down safely without risk to the form board.

8.6 Jumping Obstacles

Another method is used when the form is higher than the distance between the bottom of the backhoe and the ground. This method goes loader first over the form.

From a stockpile of dirt on the job, the operator should take a bucket full of material and dump it on the far side of the form that will be jumped, keeping it about a foot away from the inside face of the form. Another bucket full should be dumped against the form as far as your front wheels will allow.

When the extra material has been dumped, the operator should approach the form at a 90-degree angle, getting as close to the form as possible. The loader bucket should then be set on the other side of the form, resting in a flat position on the material. The operator should apply down pressure to the loader bucket until the cylinder rams bottom out and the front tires are as high as they will go.

An operator will now face one of two scenarios. Either the front wheels will be higher than the form, or the front tires will not be as high as the form.

8.6.1 High Front Tires

When the front tires are higher than the form, the operator should drive slowly forward, allowing the loader bucket to slide along on piles of dirt. As the operator approaches the form with the loader bucket in its present position, the undercarriage of the backhoe will approach the form very rapidly. As the operator drives forward with the loader bucket on the other side of the form, an operator's eyes should be on the form as the backhoe approaches it.

The front tires will be suspended over the other side of the form when the backhoe is as close to the form as possible. With the loader bucket still holding full down pressure, the operator should turn around and place the boom directly out in front, preferably with the bucket teeth stuck in the ground.

The operator should then boom down to lift the backhoe loader higher than the form. With the loader bucket on the ground it may be difficult to lift the backhoe. Should this be the case, the operator should use the outriggers and the boom to aid in the lifting effort. With the loader bucket still down, the operator should push the backhoe as it slides on the loader bucket over the form.

If, when crowd pressure is applied, the backhoe won't move, the op-

Chapter Eight: Jumping a Trench

erator can use the swing cylinders to gently shift from side to side, keeping full pressure on the crowd cylinder. As the backhoe loader crosses the form, the operator should be prepared to compensate for any drop of the machine with the boom.

8.6.2 Low Front Tires

When the front tires are not as high as the form, the operator should drive slowly forward while curling the loader bucket under in one spot. The loader bucket should not be pushed forward or backward by the movement of the backhoe. When the loader bucket is standing on the cutting edge, the operator can top. Continuing the loader past this center position will lower the wheels a couple of inches.

The front tires should be suspended on the other side of the form. If they are not, the operator should find another way around the form because it cannot be crossed without causing damage. The operator should place the backhoe bucket directly in front and lift up the backhoe higher than the form, using the outriggers for help if necessary.

8.6.3 Moving the Front Tires

The loader bucket is standing on its cutting edge, the front tires are over or on the other side of the form, and the operator cannot push the backhoe any farther because of the loader position. However, the operator still needs to get the front tires far enough from the form to lower them to the ground.

With the backhoe raised higher than the form, the operator should walk the backhoe over the form. With the backhoe teeth sunk into the ground as a solid barrier to push against, the operator should swing the backhoe from left to right using the swing cylinders, while keeping a constant eye on the crowd pressure. As the backhoe swings, the tires will get closer to the form on each side. Keeping the crowd pressure on at all times, the operator should continue swinging the backhoe back and forth a couple of feet in each direction, until the loader starts to walk forward.

Some materials are harder to walk on than others. If the walking is easy, the operator should continue until the backhoe loader can be set down safely on the other side of the form. If the walking is difficult, the operator should stop when the front loader bucket is about 4 feet from the form. The operator should then swing to one side and instead of swinging back to the other, crowd out and push the backhoe over the form on that side. When the outrigger on that side is clear of the form it should be lowered to the ground.

The operator should continue pushing with the crowd while allowing the backhoe to slide on the outrigger that is on the ground. Pushing should continue until the other outrigger is clear and it can be set down to place the boom for another bite, or until the rear tire closest to the form is clear and can be set down without damage to the form.

Chapter Nine: Cleaning Up

9.1 Introduction

Backhoe loaders do a good job of removing large quantities of material efficiently. But after all the heavy material has been moved, there is still one more job to be done: cleaning. It's true that a job should always be completed as quickly as possible, but that is no excuse for leaving a thin layer of dirt in the street or leaving a mound of dirt by the telephone pole.

> *The operator should arrange the spoils so that cleanup is quick and easy.*

Some people in the field think that backhoe loaders are not capable of fine-tuned operations such as cleaning. Consequently, laborers often have quite a bit of handwork to do after the backhoe loader's work has been completed.

But a good cleanup job takes just a few minutes. It is better than leaving a rougher grade or leaving those a few extra shovels of dirt in the street, and in the long run, those at the jobsite will appreciate the end result.

9.2 Lining Up Material with a Backstop

For example, an operator has no place to put the material but in the street. All the material needed for the project has been used, but there are still a couple of buckets full of dirt spread on the asphalt that need to be picked up. If there is a curb-and-gutter grade on the side of the street, where the operator is working, then it is just a matter of the operator scraping the material to a centralized area and using the curb and gutter as a backstop to pick up the remaining dirt.

> *When scraping material from the street, the operator should pay particular attention to the small blue reflectors epoxied to the asphalt because they mark the location of fire hydrants.*

Chapter Nine: Cleaning Up

With the loader bucket in the 3:30 position, the operator should start at the perimeter of the area to be cleaned and scrape the material in, working toward one point on the curb. The operator should try not to lay all of the material against the curb into a long stretch. If there is more than one bucket full of material in the pile, it should be lined up away from the curb and not pushed against it. Now when the operator picks up the material in the loader bucket, the weight of the material in front of the material being picked up will act as a backstop.

> *An operator should work gently around curb. One tap with the cutting edge and the curb can snap.*

9.3　Picking Up Material

For each bucket full of material picked up, the pile will widen because of the pressure placed on it by the preceding loader bite. The operator needs to make a conscious effort to keep this material contained within the confines of the loader.

The operator should line up the loader bucket perpendicular to the curb and, with the bucket in the 5 o'clock position, pull each side of the pile back and away from the curb, using the turning brakes to pull each of the two sides behind the other to make a line of material. Then the operator can take another bucket of material. The operator should continue using this method until there is about half a bucket of material left.

The remaining material should be pushed all the way up to the curb into a contained area. There should not be any material outside the edges of the loader when the material is picked up, so the operator should make sure the remaining material against the curb is contained within the outside edges of the loader. The goal is to pick up all but about a shovel's worth of material.

9.4　The Pinch

> *A loader under power can do a great deal of damage very quickly.*

At this point, with the loader bucket making contact with the curb, the operator should roll the loader bucket to the 3 o'clock or flat loader bucket position while lowering the loader bucket to place the rear loader bucket floor on the ground. The operator must roll and lower the bucket at the same time or the bucket will dig into the face of the curb.

With the feet on the brakes, the operator should slowly curl the loader bucket to the full holding position, then raise the loader bucket and back away. This is called pinching material from a curb.

An operator needs to keep a few things in mind while performing this procedure. The operator must drive forward very slowly in the last

Using the curb as a backstop.

few inches before making contact with the curb for two reasons. First, a loader under power can do a great deal of damage very rapidly. If the loader bucket is not completely on the ground, or if it is at a flatter angle than the 3:30 position, when contact is made with the curb, the loader bucket can jump up, breaking the nose of the curb when it is throttled against it. Second, if the backhoe loader is not exactly perpendicular to the curb, when the bucket edge makes its initial contact with the curb, one corner of the loader bucket will make contact first. If the operator continues with the method of pickup without correcting the loader angle, curling the bucket will gouge or break the curb on that side.

To reduce the possibility of curb damage, the operator should watch the loader bucket very closely when contact is made. If one corner touches the curb first, the operator will feel it. As the operator throttles the backhoe loader to squeeze the material against the loader edge, the operator will feel the backhoe loader start to come around slightly as it tries to line itself up to the curb. This is an indication that the loader bucket is off-center. If the operator continues pushing, the edge that made contact first will dig into the curb as the loader puts pressure against it.

9.5 Lining Up Material without a Backstop

When the operator has excess material to be picked up in the middle of the street and there are no backstops and no curb access is available, an operator needs to line up the material.

Before explaining the procedure for lining up material, an operator should know why this method works. When the loader is pushed into a large pile of material, the material fills the bucket and the pile remains in place. The pile stays in place because of adhesion, the pile's size, and the weight of the material acting as a backstop.

Lining up material is a technique that combines adhesion with the

Chapter Nine: Cleaning Up

movement in such a way that the weight created can be used to act as the backstop. The operator should start by placing the loader bucket in the 4:30 position and scraping the material toward the section designated as the lineup area. This section should be as free of obstacles as possible. The operator should pay special attention to manhole and valve covers embedded in the asphalt, because if the loader catches the raised lip of a manhole cover, it will stop a backhoe loader dead in its tracks.

> *The operator should wheelroll the outside edges of a pile, because it helps contain the material when the loader enters it.*

The operator can scrape the material into a pile and then pull the sides back into a line, or the material can be immediately scraped into a line. Either method is fine as long as it gets the material into a line 18 to 24 inches wide and no longer than about 20 feet. If there is too much material to fit into the dimensions, the remaining material should be piled at the foot of the line opposite the starting end.

The operator should then wheelroll both outside edges of the line, keeping the center material of the line loose. If there is a pile of excess material at the end of the line, the operator should wheelroll the back half of it, keeping the front half loose.

An easy way to understand what is being done is to picture that everywhere the tires touch the material, a form board goes up, hold-

Lining up material without a backstop.

ing the material in that area. If one side of the line is wheelrolled, a form board goes up, holding the material on that side. If both sides of the line and the back of the excess material pile are wheelrolled, the operator is building a form on both sides of the line and creating a backstop in back of the excess pile. As the loader is brought in to pick up the material, the forms will hold the material, forcing the material into the loader instead of pushing it along the asphalt and creating a pile in front of the loader.

To pick up the material, the operator should center the backhoe with the middle of the pile line. The loader bucket needs to be placed just over the flat position so that it is riding on the cutting edge and will enter the line at a moderate speed. Wheelrolling the outside edges of the line will force the loose center material to enter the loader bucket first, allowing the compacted outside edge material to be forced into the bucket by the compacted material directly in front of it. The operator should continue pushing the loader through the line of material until the bucket is full, until the operator makes contact with the pile behind it, or until there is no longer sufficient adhesion to hold the remaining material to the ground and it starts to slide.

When the loader bucket is full, the operator should dump it and begin again. If contact is made with the pile at the end of the line, the operator should continue pushing until either the loader bucket is full or the pile starts to slide. Once the pile starts to slide, adhesion to the ground has been broken and it becomes pointless to continue pushing. The operator should dump the material in the loader and then repeat the lineup method from the start.

When the pile is exhausted and the remaining material lined up, wheelrolled, and picked up, the backhoe loader's work is done. The rest is handwork.

9.6 Curb Shoe

Cutting grade in a street section, cutting sidewalk grade against an existing curb-and-gutter grade, and trenching behind the back of the curb all spoil dirt onto the curb-and-gutter grade. A curb shoe is a device that lies on top of the gutter and pushes the unwanted dirt into the street as it is pulled along. The curb shoe is designed to fit on the outside edge of a loader grader's blade, but it can be easily modified to accommodate the teeth of a backhoe bucket to serve the same purpose.

Taking the time to make or modify a curb shoe to fit a backhoe is time well-spent when the operator is going to cut sidewalk grade for long periods of time. But when a backhoe must be moved from item to item on a jobsite, dragging around a curb shoe that will be used only once or twice is not feasible.

Chapter Nine: Cleaning Up

9.7 Cleaning Curb-and-gutter

For example, an operator is about to cut a 50-foot stretch of sidewalk grade against existing curb-and-gutter, and a curb shoe is not available because the operator was told only removals were being done. The operator knows that the dirt will spoil on top of the curb and into the gutter and that two cleaning passes will have to be made. If the gutter was cleaned first, then cleaning the top of the curb after that would spoil the dirt down into the gutter just cleaned.

Before an operator starts cleaning, he or she should take a good look at the materials being used. If wood is being used, the operator should make sure that it is thick enough to withstand the pressure from the teeth and it is in good enough condition that it won't split in the middle of a pass. Most operators should use 4x4s or larger.

For cleaning curb-and-gutter after cutting sidewalk grade, three pieces of wood work well: one piece to ride on top of the curb and two pieces to ride in the gutter. The first piece in the gutter is to remove the heavy material and the second piece is for cleaning any leftovers.

If an operator is cutting sidewalk grade, there is no need to reposition the backhoe to clean the curb after the last pass of the sidewalk grade. The operator can cut the grade and clean the curb-and-gutter from the same setup position. After the grade has been cut, the operator should reach back very carefully to the piece of wood positioned on top of the curb and slowly pull it in. It takes practice, but the operator must maintain equal pressure on the top of the curb while drawing in at a slightly offset angle. The piece of wood must be drawn straight and with the right amount of pressure on the top of the curb. The backhoe is positioned in the sidewalk area, reaching over to the curb area. If the teeth slip off the front or the side of the wood, the curb will be chipped, so the operator needs to be careful.

A small rock between the wood and concrete will scratch both materials.

The next piece on the gutter is a little more difficult because the farther out to the side the operator reaches with the bucket, the greater the angle the operator will need to compensate for as the backhoe is drawn in. The piece of wood should be set at a slight angle to the curb face. When the dirt starts to pile up in front of it, the operator will want it to be pushed off the lip of the gutter and not back into the curb face.

With the teeth positioned lengthwise so that they make full contact with the wood, the operator should pull the piece along very slowly. If there is an excessive amount of dirt in the gutter, the dirt will pile up against the board and the operator may lose sight of the teeth. If this happens, the operator should stop, reach back and grab the second piece of wood and, at the same angle, drag it toward the backhoe, cleaning the remaining debris, if any, from the gutter. When the operator can again see the board, the operator should continue pulling the rear board, letting it push the front board along until the operator matches the point of the sidewalk grade. The operator should then push back, cut the next stretch of sidewalk grade, and repeat the curb cleaning process again.

9.8 Cleaning Sidewalk

An old barricade, a long 4x4, or an old pallet will work well for cleaning sidewalk. In this example, an operator has just completed backfilling a jacking pit located lengthwise in the middle of a 5-foot sidewalk and there is loose material to clean up.

Backhoe setup is really not an issue as long as the operator can pull straight into the area where the dirt needs to go. Using an old barricade from the trash pile, the operator should set the barricade at the far end of the area to be cleaned, and then set the teeth on the barricade end closest to the operator. If the operator set the teeth on the far end of the barricade, material would accumulate under the front as it was pulled in, lifting the barricade and allowing it to ride on top of the dirt.

A barricade has two different directions of wood: two pieces for the cross braces running one way, and two pieces for the legs running the opposite way. Since the cross braces are nailed to the outside of the legs, they ride flat on the sidewalk when placed in the correct position. The operator should pull the barricade in, monitoring the amount of pressure being used. The closer the barricade gets to the operator, the faster the operator will have to raise the boom to compensate for the increase in crowd pressure.

A piece of wood works well for this method of cleanup, especially if there is not a great deal of material left. As the material starts to build against the piece of wood as it is drawn in, the operator can lose sight of the teeth very rapidly. Therefore, an operator would be wise to save the wood for light cleanup, or combine the two tools by first using a barricade or pallet and then finishing with a piece of wood.

The piece of wood must be at least as long as the sidewalk is wide, otherwise two passes will have to be made. The operator should set the wood out far enough to retrieve the material and then pull it in, dragging the wood against the sidewalk into the jacking pit.

When using wood and repositioning the teeth, an operator should take a 4x4 longer than the width of the bucket and place it on the ground in front of the backhoes that it's running in the same direction as the teeth. With the backhoe, the operator should reach out and set the teeth on the piece of wood, making sure the teeth are running down the middle of the wood. The operator should apply the boom pressure only, forcing the teeth into the wood. Either the teeth will split the wood under the pressure, or the teeth will start to sink in-

Chapter Nine: Cleaning Up

to the wood, and as the pressure increases, the piece of wood will roll, leaving the teeth to slam against the surface on which the wood is resting. The easy remedy to this is for the operator to offset the angle of the wood slightly in relationship to the bucket teeth. Instead of the teeth riding right down the middle of the wood, they will be running at a slight diagonal across the wood.

9.8.1 Picking Up Material

It is easier to pick up compacted material with the loader bucket than loose material, so the operator needs to wheelroll it. The operator should position the backhoe parallel to the curb, with one set of tires in the gutter. Rolling over the material until it is nice and compact, the operator can pinch the curb to get the rest of the material. The operator must approach the curb at exactly a 90-degree angle, and when the loader edge meets the face of the curb, it must lie perfectly flat against it. A good way to check the angle before contact is made with the curb is for the operator to hold back from the curb about 5 or 6 feet, stop, roll the loader bucket over to the 4 o'clock position and, from the line of sight at the operator's seat, match the top edge of the loader bucket to the top edge of the curb. If one side of the bucket is higher than the other side, the operator should bring the loader straight by slightly changing the angle of approach to the curb. This technique of leveling up to the curb will work anywhere there is an asphalt street sided in curb-and-guttter grade.

When the operator is sure the top of the loader bucket is level to the curb, the operator should approach it with the loader bucket in the 3:30 position, lower the bucket to the ground, and slowly push into the base of the curb, allowing the material to fill the bucket. When the operator thinks contact has been made with the curb, the backhoe loader should be throttled up for just a second to squeeze the material onto the loader edge. The operator should then back off the throttle completely. Now the operator can either push in the clutch or take the backhoe loader out of gear, letting the machine relax back from the curb.

When the pressure is let off the loader bucket, the machine will roll back slightly and the operator should apply the brakes hard. The loader bucket still needs to be touching the curb but not applying pressure to it. The operator can then roll the bucket to a full holding position and the material can be dumped.

Chapter Ten:
Moving a Stuck Backhoe Loader

10.1 Introduction

When a backhoe loader *really* gets stuck, it's stuck and nothing but a crane or dozer is going to get it unstuck. But when an operator refers to getting a backhoe loader stuck, he or she is usually referring to the fact that somehow he or she maneuvered the backhoe loader into a position in which he or she does not have the knowledge to get it out of.

An operator gets a backhoe loader stuck for one of two reasons. First, the rear tires are unable to make contact with the ground due to a high centering effect of the backhoe. This commonly occurs when, wheel-rolling a trench, the side gives way. The front and rear tires on the trench side fall into the trench, placing the backhoe loader in an off-balance position where it is unable to gain rear traction. Second, the working material is very soft or wet and has built up in front of the tires, creating enough resistance to the front tires or chassis of the backhoe loader to permit free rear wheel spin. This is very common when working in sand or when trying to mix dry material with wet material in the hope of drying it out. The common denominator in both of these situations is lack of traction to the rear tires. If the backhoe loader cannot be driven out of a situation, an operator must turn around and use the backhoe.

There are four ways to get a backhoe loader unstuck. They are listed here from easiest to most difficult, and they should be attempted in that order. An operator may find that more than one of these methods is necessary.

10.2 Loader Walk

Loader walking can be used in every situation from backing up an incline to backing out of slippery material. To begin, the operator should set the loader bucket in the 6 o'clock position, raising the front tires clear of the grade. With the transmission set in reverse, the operator should allow the rear tires to spin freely, curl the loader bucket, and push back. When the loader bucket has uncurled and the reverse motion has stopped, the operator should step on the brakes, disengage the reverser, lift the loader bucket, and set it back to the 6 o'clock position for another bite. The operator should then set the loader bucket back into the grade, lifting the front tires clear, engage the reverser, and release the brakes, repeating the process.

10.3 Drive-out

The term drive-out refers to engaging the transmission and allowing the rear wheels to spin, thus aiding the operator in the push/pull method. When clay-type material is sprinkled with water, it will make the surface wet enough to allow serious tire spin. In this instance, driving out is necessary.

The operator needs to keep three things in mind. First, the bucket will be used in both methods and so the teeth will probably penetrate the ground. The operator needs to be sure they do not penetrate the surface near any underground utilities. Second, there are no front-tire steering capabilities in slippery material. The operator will steer the backhoe loader by either using the steering brakes or by pointing the machine in the direction to be traveled, pushed or pulling in that direction. Third, even though steering capabilities are gone, the operator must continue to keep the front tires straight. When turned, the front tires or rims can become damaged when being pulled in a straight direction, and they can produce quite a bit of drag to the backhoe loader as well.

> **SAFETY TIP**
> An operator should remember a backhoe loader could regain traction and move under its own power at any time.

10.4 Pushout Method

When using the pushout method, the operator should, without turning the seat around, reach back with one or both hands, open or extend the backhoe bucket as far as it will go, and prepare to set the boom to the ground. With the loader bucket just a foot off the ground, the operator's feet on the steering brakes and the backhoe loader still in the forward gear, the operator should bring the fully open backhoe bucket to a point halfway in, sink the teeth into the ground using the crowd and the boom to push, and allow the rear tires to spin. The rear tires should remain on the ground with help from the boom-and-crowd.

When the crowd-and-boom have run out, the operator should disengage the transmission, bring the crowd-and-boom back in, re-engage the transmission, and start again. This process should be continued until the operator no longer requires the help of the backhoe hydraulics.

The operator must be alert to the fact that the backhoe loader could regain traction and start to move under its own power at any time. If the sudden movement of the backhoe catches the operator by surprise, the operator should boom down, raising the machine just enough to take the weight off the rear tires.

A drawback to the pushout method is the tendency for wet or loose material to build up on and around the front of the front axle and tires. When the material is soft enough to permit excessive buildup, the backhoe hydraulics can no longer overcome the drag created by the mud and the backhoe loader will come to a stop.

> **SAFETY TIP**
> An operator should watch out for shallow underground utilities.

10.5 Pivoting

A variation of the pushout method is the pivot. The pushout method pushes the backhoe loader straight forward, while the pivot method frees the machine by pushing in a sideways direction.

For example, a side of the trench being wheelrolled collapses. The rear and front tires are wedged in the trench, leaving the backhoe loader on its side in the trench. Both the pushout and pullout method will drag the machine along the trench, destroying it. The walkout method will do even more damage.

Making sure the safety belt is fastened, the operator should place the loader bucket in the flat position, elevate it just above the trench, and slowly move the boom and place it on the side of the trench that the tires are on. If the left side of the backhoe is in the trench, the boom should be placed on the left side of the trench. The boom should be placed far enough from the side of the trench that the loader will not cave it in when pressure is applied to the boom.

The operator should boom straight down until the backhoe loader is level and then set the loader bucket down to the surface with enough pressure to lift the front tires off the trench. Using the crowd and the boom, the operator should push the backhoe until the rear tire closest to the trench is on stable ground. The operator should then fold up the backhoe and, with the loader bucket still maintaining the front wheels elevated, back off the trench.

The pivot method can be used for a variety of applications. The important thing to remember is that if the operator does not place weight on the loader as it is pushed, the front half of the backhoe will slide opposite the direction in which the operator is pushing. In the case described above, that would send the front end of the backhoe loader completely into the trench.

Chapter Ten: Moving a Stuck Backhoe Loader

10.6 Pullout Method

The pullout method is the opposite of the pushout method with two exceptions. First, the operator will lose all front tire steering control when working in wet material. The only way to maintain any direction of movement is to pull the backhoe straight in the direction desired. Second, the teeth will penetrate the ground. Depending on the type of material the backhoe loader is in, this penetration could be 2 or more feet. The operator should be certain there are no utilities in the area.

Since steering is not an option in the pullout method, the operator can either sit facing forward or facing the backhoe, whichever method works best. The loader bucket should be set about a foot above the material and kept there until the procedure is completed.

With the front tires turned straight to reduce drag, the backhoe stretched out three-quarters of its reach, and the bucket in the full open position, the operator should sink the teeth into the grade. The boom pressure will control the amount of weight placed on the rear tires.

The boom-and-crowd are in their positions to help the tires move the backhoe. As the operator pulls in, he or she should concentrate on keeping the rear of the backhoe loader on the ground and the rear tires in contact with the surface. With the transmission in low gear and the idle control lever set around 1,500 r.p.m., the operator should engage the transmission and allow the rear tires to spin in reverse. The boom-and-crowd should be pulled in, allowing the backhoe loader to move in the direction desired. When the machine starts to move, the boom alone should be used to regulate the traction to the rear tires. Increased pressure to the boom will allow the tires to spin more freely, whereas decreased pressure will apply more weight to the rear, causing the tires to bite harder into the grade. If the sudden movement of the backhoe loader catches the operator by surprise, the operator should boom down to raise the rear tires from the grade, thereby stopping all backhoe movement.

10.7 Walkout Method

Resistance to the chassis as it pertains to wet material is probably the most difficult obstacle to overcome when dealing with wet material. Water, as it puddles on or is mixed with dirt, creates mud and has the unique ability to displace air when a backhoe loader chassis is placed on it, creating a vacuum or suction. This suction, combined with the chassis in the mud surrounding four tires, creates a great deal of force holding the backhoe loader to the ground.

The operator knows the rear tires are unable to make traction in the wet material. But if the tires were able to make traction, could the operator drive the backhoe out? If the operator looks at the size of the area in which the backhoe loader is located, the operator will realize that as the backhoe loader starts to move under its own power, the wet material will build up in front of the front tires. As the rear tires start to move, the wet material will start to build up in front of them as well.

The material in front of the tires will produce enough resistance to the machine to permit rear tire spin. In wet material, the spinning rear tires will act as two shovels burying the backhoe loader even deeper into the mud. Driving out is now no longer an option. The operator will have to use the walkout method.

The loader bucket is set at the 6 o'clock, or full dump, position, and the boom-and-crowd are set ¾ the way out, the backhoe opened, and teeth planted firmly in the ground. With one hand the operator will work the loader control lever and the other hand will work the backhoe functions.

> **SAFETY TIP**
> When in the backhoe loader an operator should always wear the safety belt.

The operator should lift up the rear of the machine using the boom until it is clear of the grade. Lowering the loader bucket until the cylinders bottom out and the front tires are as high above the grade as possible, the operator should pull straight in, placing the backhoe in a loaded position. Because the loader bucket is in the ground when the boom-and-crowd are pulled in, the backhoe will not move. The pressure administered by pulling in on the boom-and-crowd is still present in the cylinders.

The operator needs to execute both sets of functions at the same time and at the same speed. If the boom-and-crowd are not pulled in at the same time with the teeth planted in the ground, the loader will curl in one spot, digging itself a hole instead of pushing the backhoe.

Pulling in on the boom-and-crowd with the loader bucket stationary in the 6 o'clock position will produce the same results. The backhoe will dig itself a hole while the rest of the machine stays in one place.

Now that the backhoe bucket is loaded and the loader bucket is off the ground, the operator should curl the loader bucket while simultaneously reaching back and pulling in on the backhoe control levers. When the backhoe walks back a few feet, the operator will run out of curl on the bucket.

Reaching back with the backhoe to take another bite, the operator should then reset the loader bucket to the 6 o'clock position and continue this walking until the backhoe loader is clear.

Index

Accidents, 17-23
American standard system, 29
Asphalt (*see* Removals)
Backfilling, 75-86
    Trenches, 78
    Gauging material flow, 79-80
    At 45 degrees, 84-86
    At 90 degrees, 82-84
    Sidecutting, 80-82
Backhoe accessories, 4-7
    Auger, 4
    Breaker, 4-5
    Vibratory plate, 5
    Impact tamp, 5
    Sheepsfoot, 5-6
    Quick-change bucket assembly, 6
    4-in-1/Clamshell loader, 6-7
    Removable loader teeth, 7
    Asphalt cutter, 7
    Lifting forks, 7
Backhoe bucket, 12-13
    Changing, 23-24
    Maintenance, 12, 13
    Bending, 13
    Power positions, 13

Backhoe maintenance, 7
    Warmup, 7
    Oil, 8
    Grease, 8
Backhoe maneuvering
    Loader walk, 153-154
    Drive-out, 154
    Pushout, 154-155
    Pivoting, 155
    Pullout, 157
    Walkout, 157
Backhoe selection, 1
    Center mount, 2
    Offset, 2-3
    Size, 3-4
Backstops, 122, 124, 133, 145, 152
Banjo, 31, 32
Benching, 21-22
Boring, 105
Bridging, 84
Caution tape, 90
Cave-ins, 19-20
Cavitation, 14
Chains, proper use, 12

Index

Cleanup
    Lining up material, 145-146
    Picking up material, 146, 152

Communication, 25-29
    Hand signals, 25-29
    Cut signals, 25
    Fill signals, 25

Compaction, 75-76
    Adding water, 76
    Trenches (*see* Backfilling)
    Optimum water content, 76

Controlled relief (*see* Removals)

Crimping, 111-112

Curb and gutter
    Digging under, 52-53
    Setup, 39
    Hike-down, 39
    Hike up, 31
    Cleaning, 150
    Grade, 31-32, 38-42

Curb bench, 40

Curb shoe, 149

Cut sheet, 30

Cycling, 45-46

Digouts, 127-129
    (*see also* Removals)

Direct burial (*see* Telephone lines)

Electrical conduit
    Locating, 98-101
    Types, 98
    Breaking lines, 109-110
    Making contact with, 108
    Surges, 110
    Sweeps, 100
    Transmission, 108-109
    High voltage, 110

Eye protection, 24

Fiber optic (*see* Telephone lines)

Footings, 53-58

Gas lines
    Locating, 105-106
    Backfill sand, 105
    Setup, 106
    Making contact with, 106, 112-113, 113-114

Grade stakes, 29-33
    Abbreviations, 42-43

Grading
    Curb and gutter, 31-32, 38-42
    Import, 74-75
    Loader, 66-75
    Sidewalk, 32, 33-38
    To a raised perimeter, 68

Guide pass, 94

Hand signals (*see* Communication)

High voltage (*see* Electrical conduit)

Jumping a trench
    At 45 degrees, 136-137
    Back first, 140-141
    On the go, 135-136
    Over obstacles, 142-143
    Straight, 137-139

Keyway, 35, 121

Lifting, 75, 77-78

Lining up material (*see* Cleanup)

Load, leveling, 65

Loader
    Positions, 68-74
    Grading, 66

Loader bucket, 9-12
    Edges, 66
    Picking up pieces from a flat surface, 131
    Picking up pieces out of the grade, 130, 133

Loading (*see* Trucks)

Location services (*see* Utilities)

Offsets, 30, 31

Index

Optimum water content (*see* Compaction)
Outriggers
  Hike-up, 34, 40
  Placement, 35-36
Parallel setup, 94
Perpendicular setup, 95
Picking up pieces
  Out of the grade, 124
  Uneven pieces, 126-127
  From a flat surface, 122-123
Piers, 58-60
Pinch, the, 146-147
Pins, 8, 9, 47
  Gaulding, 8, 9
  Hot spots, 8, 9
Pull box, 88, 98-99
Pushing back, 48-49
Receiving hole, 51, 52
Removals, 117, 129-130
  Techniques, 117-118
  Saving edges, 118-122
  Controlled relief, 118
Roading, 23
Safety belts, 24
Sawcuts, 127
Scribing, 128
Sensitivity zone, 96
Sidecutting (*see* Backfilling)
Sidewalk
  Grade, 32, 33-38
  Hike-up, 34
  Excess material, 34
Sloping, 20-21
Stepping, 20, 21-22, 24
Signals (*see* Communication)

Subgrade
  Curb and gutter, 32
  Sidewalk, 32
Station marks, 30
Steering brakes, 15, 154
Swing radius, 18
Telephone lines
  Direct burial, 115
  Locating, 107
  Making contact with, 114-115
  Fiber optic, 107
Tracer wire, 106
Transformers, 99, 109
Transite (*see* Water lines)
Trenching, 45-47
  Corrections, 47
  Straight-line, 47
  (*see also* Cycling)
Trench signs, 108
Trucks
  Spotting, 62-63
  Loading, 63-65
  Bedding, 66
Utilities
  Digging under, 49-52
  Marks, 88, 89-90
  Poles, 99-100
  Locating, 87-98
  Location services, 89-90
Vaults, 98-99
Water lines
  Locating, 101-105
  Plastic, 103-104
  Steel, 104-105
  Transite, 102-103
Wheelrolling, 77, 78-79